TEXTBOOK
OF
PHYCOLOGY

Dr. Pooja
Dept. of Botany
R.C.C. College
Ghaziabad (U.P.)
(India)

DISCOVERY PUBLISHING HOUSE PVT. LTD.
NEW DELHI-110 002

First Published - 2010

Reprinted - 2015

ISBN: 978-81-8356-582-0

Textbook of Phycology

Published by:

DISCOVERY PUBLISHING HOUSE PVT. LTD.

4383/4B, Ansari Road, Darya Ganj

New Delhi-110 002 (India)

Phone: +91-11-23279245, 43596064-65

Fax: +91-11-23253475

E-mail: discoverypublishinghouse@gmail.com

sales@discoverypublishinggroup.com

web: www.discoverypublishinggroup.com

Printed at:

Infinity Imaging Systems

Delhi

Preface

Phycology includes the study of prokaryotic forms known as blue-green algae or cyanobacteria. A number of microscopic algae also occur as symbionts in lichens.

The development of the study of phycology runs in a pattern comparable with, and parallel to, other biological fields but at a different rate. After the invention of the printing-press in the 15th century (with the publication of the first printed book: Gutenberg's Bible of 1488) education enabled people to read and knowledge to spread.

Phyology is the study of algae. Human interest in plants as food goes back into the origins of the species (Homo sapiens) and knowledge of algae can be traced back more than two thousand years. However only in the last three hundred years has that knowledge developed into a rapidly developing science.

The study of botany goes back into pre-history as plants were the food of people from the beginning of the human race. The first attempts at plant cultivation are believed to have been made shortly before 10,000 BC in Western Asia (Morton, 1981) and the first references to algae are to be found in early Chinese literature. Records as far back as 3,000 BC indicate that algae were used by the emperor of China as food The use of Porphyra in China dates back to at least A.D. 533-44, there are also references in Roman and Greek literature. The Greek word for algae was "Phycos" whilst in Roman times the name became Fucus. There are early references to the use of algae for manure. The first coralline algae to be recognized as living organisms were probably Corallina, by Pliny the Elder in the first century AD.

Algae (*sing*. alga) are a large and diverse group of simple, typically autotrophic organisms, ranging from unicellular to multicellular forms.

There are nearly 30,000 algae species. The largest and most complex marine forms are called seaweeds, with 10,000 species. They are photosynthetic, like plants, and "simple" because they lack the many distinct organs found in land plants. Though the prokaryotic *cyanobacteria* (commonly referred to as blue-green algae) were traditionally included as "algae" in older textbooks, many modern sources regard this as outdated and restrict the term *algae* to eukaryotic organisms.

–Author

Contents

Chapter–1

INTRODUCTION

Phycology (from, phykos, "seaweed"; and , -logia) or algology (from Latin alga, also "seaweed"), a subdiscipline of botany, is the scientific study of algae. Algae are important as primary producers in aquatic ecosystems. Most algae are eukaryotic, photosynthetic organisms that live in a wet environment. They are distinguished from the higher plants by a lack of true roots, stems or leaves. Many species are single-celled and microscopic (including phytoplankton and other microalgae); many others are multicellular to one degree or another, some of these growing to large size (for example, seaweeds such as kelp and Sargassum).

Phycology also includes the study of prokaryotic forms known as blue-green algae or cyanobacteria. A number of microscopic algae also occur as symbionts in lichens.

A phycologist is a person who studies algae as described above. In a similar manner, a mycologist is a person who has been professionally trained in mycology, the study of fungi. Phyology is the study of algae and history is the study of the past human activities. Human interest in plants as food goes back into the origins of the species (Homo sapiens) and knowledge of algae can be traced back more than two thousand years. However only in the last three hundred years has that knowledge developed into a rapidly developing science.

The study of botany goes back into pre-history as plants were the food of people from the beginning of the human race. The first attempts at plant cultivation are believed to have been made shortly before 10,000 BC in Western Asia and the first references to algae are to be found in early Chinese literature. Records as far back as 3,000 BC indicate that algae were used by the emperor of China as food The use of Porphyra in China dates back to at least A.D. 533-44, there are also references in Roman and Greek literature. The Greek word for algae was "Phycos" whilst in Roman times the name became Fucus. There are early references to the use of algae for manure. The first coralline algae to be recognized as living organisms were probably Corallina, by Pliny the Elder in the first century AD.

The classification of plants suffered many changes since Theophrastus (372-287 B.C.) and Aristotle (384-322 B.C.) grouped them as "trees", "shrubs" and "herbs".

Little is known of botany during the Middle Ages — it was the Dark Ages of botany.

The development of the study of phycology runs in a pattern comparable with, and parallel to, other biological fields but at a different rate. After the invention of the printing-press in the 15th century (with the publication of the first printed book: Gutenberg's Bible of 1488) education enabled people to read and knowledge to spread.

Written accounts of the algae of South Africa were made by the Portuguese explorers of the 15th and 16th centuries, however it is not clear to which species reference was being made.

In the 17th century there was a great awakening of scientific interest all over Europe, and after the invention of the printing-press books on botany were published. Among them was the work of John Ray who wrote in 1660: Catalogus Plantarum circa Cantabrigiam., this initiated a new era in the study of Botany Ray "influenced both the theory and the practice of botany more decisively than any other single person in the latter half of the seventeenth century." However, no real progress was made in the scientific study of algae until the invention of the microscope—in about 1600. It was Anton van Leeuwenhoek who discovered bacteria and saw the cell structure of plants. His unsystematic glimpses of plant structure, reported to the Royal Society between 1678 and his death in 1723, produced no significant advances

As adventurers explored the world more species of all animals and plants were discovered, this demanded efforts to bring order out of this quickly accumulating knowledge.

Australia The first Australian marine plant recorded in print was collected from Shark Bay on the Western Australian coast by William Dampier who described many new species of Australian wildlife in the 17th century

Before Carl von Linné (1707-1778) animals and plants had names, but it took him to arrange the names and group the plants of this Earth in some sort of order. Carolus Linnaeus (Carl von Linné) was a Swedish botanist, the son of a pastor of the Lurtheran church, a physician and zoologist. He laid the foundations of modern biological systematics and nomenclature in his Species Plantarum (1753). He adopted and popularized a binomial (or binary) system of designation (Morton, 1981) using one name as the genus and a second name as the species name both in Latin or Latinised. This specific name he referred to as a trivial name nomen triviale consisting of a single word, normally a Latin adjective, but any single word would suffice, to identify a particular species, but not intended to describe it. He developed a coherent system for naming organisms and divided the plant kingdom into 25 classes one of which, the Cryptogamia, included all plants with concealed reproductive organs. He divided the Cryptogamia into four orders: Filices, Musci (mosses), Algae — which included lichens and liverworts and fungi.

Examination for the reproductive structures had already started. In 1711, R.A.F de Réaumur gave an account of Fucus in which noted the two types of external openings in the thallus: the non-sexual cryptostromata (sterile surface cavities) and the conceptacles (fertile cavities, immersed but with a surface opening) containing the sexual organs, which he thought were female flowers. With a lens he was able to see the oogonoa (the female sex organs) and the antheridia (the male sex organs) within the conceptacles, but he interpreted these as seeds. Johann Hedwig (1730-1799) provided further evidence of the sexual process in algae, and figured conjugation in Spirogyra Hedwig in 1797. He also illustrated Chara and identified the antheridia and oogonia as male and female sexual organs.

Harvey commented on...motion, apparently spontaneous, among the seeds at the period of germination. Some found it difficult...to account for these anomalous motions...that the seeds becomes (how is not said) a perfect nimalcule, which after enjoying an animal existence for a time ceases to live animally, and, reverting to its original nature, gives birth to a vegetable. Thus, this seed was first vegetable, then animal, and then again vegeable,... During the 18th Century there was a stormy controversy as to whether coralline algae were plants or animals. Up to the mid-1700s coralline algae (and coral animals) were generally treated as plants. By 1768 many, but by no means all authorities, considered them animal. Five years later, Harvey concluded that they were certainly of vegetable material he noted:

> "The question of the vegetable nature of Corallines, among which the Melobesia take rank, may now be considered as finally set at rest, by the researches of Kützing, Phillipi and Decaisne."

The first scientific species description of a South African seaweed accepted for most nomenclatural purposes is that of Ecklonia maxima, published in 1757 as Fucus maximus.

Knowledge of North American Pacific algae begins with the 1791-95 expedition of Captain George Vancouver.

Archibald Menzies (1754-1842) was the appointed botanist on the expedition led by Captain George Vancouver in the ships Discovery and Chatham of 1791–1795 to the Pacific coast of North America and south-western Australia. The algae collected by Menzies were passed to Dawson Turner (1775-1858) who described and illustrated them in a four-volumed work published in 1808–1819. However Turner only referred to the taxa referable to Fucus; either Menzies collected very few or he gave only a few to Turner. Three of these species described by Turner later became the types of new genera (Papenfuss, 1976) and (Huisman, 2000) Turner also received plants from Robert Brown (1773-1858) the botanist who accompanied Captain Matthew Flinders on the Investigator (1801-1805). This collection also included many plants from Australia

The real awakening of interest in American algae resulted from a visit by William Henry Harvey in 1849-1850 when he visited areas from Florida to Nova Scotia and produced three volumes of Nereis Boreali-Americana. These gave an incentive to others to study algae.

The first collector of marine algae in Greenland waters seems to have been J.M.Vahl who lived in Greenland from 1828 to 1836. Vahl's East Greenland species were not recorded until 1893 when Rosenvinge included them in his work of 1893 together with the species collected by Sylow F.R.Kjellman records only 12 species from East Greenland 4 of which are doubtful, these records are based on Zeller's list.

Carl Adolph Agardh was one of the most prominent algologists of all time, he was born in Sweden on 23 January 1785 and died on 28 January 1859. He was Professor of Botany at the University of Lund and later Bishop of Karlstad Diocese (Papenfuss, 1976) Many species still show his name as the authority of the scientific name. He traveled widely in Europe visiting Germany, Poland, Denmark, the Netherlands, Belgium, France and Italy and was the first to emphasize the importance of the reproductive characters of algae and use them to distinguish the different genera and families. His son, Jacob Georg Agardh (1813-1901), who became Professor of Botany at Lund in 1839, made a study of the life-histories of algae, described many new genera and species. It was to him that many workers sent specimens for determination and as donations. Because of this the herbarium at Lund is the most important algal herbaria in the world.

The first records of algae from the Faroe Islands were made by Jørgen Landt in his book of 1800 where he mentions about 30 species. Following this, Hans Christian Lyngbye visited the Faroe Islands in 1817 and published his work in 1819. In this, he described several new genera and species, some 100 new species were listed. Emil Rostrup who visited the Faroe Islands in 1867 listed ten new species and a total not far from 100. In 1895, Herman G. Simmons mentioned 125 species. In that year F. Børgesen started work and in 1902 published his work.

Jean Vincent Félix Lamouroux was the first, in 1813, to separate the algae into groups on the basis of colour (Dixon and Irvine, 1977 p.59). At this time all coralline algae were considered animals, it was R. Philippi who in 1837 published his paper in which he finally recognized that coralline algae were not animals and he proposed the generic names Lithophyllum and Lithothamnion.

Freshwater algae are commonly treated separately from marine algae and may be considered not correctly placed in phycology. Lewis Weston Dillwyn (1778-1855) "British Confervae" (1809) was one of the earliest attempts to bring together all that was then known on the British Freshwater algae.

Specimens of Anne E. Ball (1808-1872) have been found in both the Herbaria of the Irish National Botanic Gardens, Dublin and the Ulster Museum (BEL). A.E.Ball was an Irish algologist who corresponded with W.H. Harvey and whose records appear in his Phycologia Britannica. The specimens in Dublin do not contain any unusual or rare items. However, they are well documented.

Willian Henry Harvey (1811-1866), Keeper of the Herbarium and Professor in Botany at Trinity College, Dublin, was one of the most distinguished algologists of his time. Apart from Ireland he visited South Africa, the Atlantic seaboard of

America as far south as the Florida Keys on the east coast of North America and Australia (1854-1856). Between 1853 to 1856 he visited Ceylon, Australia and New Zealand and various parts of the South Pacific His collection in Australia resulted in one of the most extensive collections of marine plants and it inspired others. He published: *Nereis Australis* Or *Algae of the Southern Ocean* in 1847-1849 and in 1846-51 his *Phycologia Britannica* appeared. His *Nereis Boreali-Americana* was published in three parts (1852-1858) this was the first, and still is (1976) is the only marine algal flora of North America as it includes taxa from the Pacific coast . His five-volume *Phycologia Australica* was published in 1858 to 1863. These volumes remain to this day a most important reference to Australian algae. His primary herbarium is in Trinity College, Dublin (TCD). However large collections of Harvey material are to be found in the Ulster Museum (BEL) University of St Andrews (STA) and National Herbarium of Victoria (MEL), Melbourne, Australia (May, 1977). Many of the collectors of this period sent, and exchanged, specimens freely one to another, as a result Harvey's books show a remarkable knowledge of the distribution of algae elsewhere in the world. His Phycologia Britannica lists species recorded and collected from various parts of the British Isles. For example he notes William Thompson (1805–1852), William McCalla (c.1814-1849), John Templeton (1766-1825) and D. Landsborough (1779-1854) who collected, as he did, from distinct sites in Ireland. The collections of these botanists, and many others, are represented separately by collections in the Ulster Museum.

Sir William Jackson Hooker (1785-1865) was a life-long friend of Harvey he was appointed Professor of Botany at Glasgow University in 1820 and became Director in Kew 1841-1865. Hooker recognized the talent in Harvey and lent him books, encouraged and invited him to write the section on algae in his British Flora. as well as the section on algae for The Botany of Captain Beechey's Voyage. Margaret Gatty (1809-1873) (née Margaret Scott) (author of British Seaweeds, 1863), and others, corresponded with William Henry Harvey (Desmond, 1977 and Evans, 2003).

Much work was done in this period by many workers and the many specimens became very valuable. Harvey's specimens, are to be found in at least several herbaria as well as those of other phycologists whose names are to be found in historic publications. In the same period Friedrich Traugott Kützing (1807-1893) in Germany described more new genera than anyone either before or after His publications span the period 1841 to 1869 and added materially to knowledge of algae of cold waters of the Arctic seas. Some of his specimend are stored in the Ulster Museum Herbarium (BEL) catalogued: F1171; F10281–F10318. In 1883 Frans Reinhold Kjellman, Professor of Botany at Uppsala University, published The Algae of the Arctic Sea. He divided the "Arctic Sea" into different regions which surround the North Pole Further research work on the marine algae of the world included: Charles Lewis Anderson who collaborated with William Gilson Farlow and with Professor Daniel Cady Eaton to produce on the first exsiccatae of North American Algae (Papenfuss, 1976) Edward Morell Holmes (1843-1930), was an expert on seaweeds, mosses, liverworts and lichens, specimens were sent to him from all over the British Isles, as well as from Norway, Sweden, Florida, Tasmania, France, Cape of Good Hope, Cylon and Australia. He also exchanged specimens . and some are in the herbarium of the Ulster Museum (BEL). George Clifton an Australian phycologist is mentioned in Harvey's Memoirs, as the Superintendent of the Water Police in Perth, West Australia sent algal specimens to Harvey. In these years there were many workers in this field: W.G. Farlow, mentioned above, who was appointed in 1879 Professor of Cryptogamic Botany at University of Harvard in 1879 and published, among other works, the Marine algae of New England and Adjacent Coasts.; in 1876 John Erhard Areschoug, a Swedish Professor of Botany at Upsalla University, reported on some brown algae collected in California by Gustavus A. Eisen. George W. Traill (1836-1897) was a clerk in the Standard Life Company in Edinburgh where he worked long hours, yet he was one of the greatest authorities on Scottish algae. Despite bad health he was an indefatigable collector. In 1892 he gave his collection to the Herbarium of the Edinburgh Botanic Gardens (Furley, 1989).

Mikael Heggelund Foslie (M.Foslie) (1855-1905) published 69 papers between 1887-1909. During this time he increased the number of species and forms (of corallines) from 175 to 650. After his death his collection of specimens was purchased by the Museum of the Royal Norwegian Society for Sciences and Letters and there is a small collection of his in the Ulster Museum Herbarium: entitled: *Algae Norvegicae* (Ulster Museum Herbarium catalogue (BEL): F10319–F10334). F.Heydrich also described 84 taxa and was a bitter foe of Foslie. This left a legacy of complicated and still unresolved problem

It was in the 19th Century that the true nature of lichens, as organisms consisting of an alga and a fungus in specific association, was demonstrated by Schwendener in 1867. This removed a source of confusion in morphology and classification (Morton, 1981 p.432 It was in this period (1859) that Charles Darwin (1809-1882) published his book on evolution, *On the Origin of Species by Means of Natural Selection*.

In 1895 Børgesen started his study of the Faeroe Islands and published his work in 1902 Later between 1920 and 1936 he published his research on the algae of the Canary Islands.

In 1935 and 1945 Felix Eugen Fritsch (1879-1954) published in two volumes his treatise: The Structure and Reproduction of the Algae. These two volumes detail virtually all that was then known about the morphology and reproduction of the algae. However knowledge of algae has so greatly increased since then it would be impossible for these to be brought up-to-date. Nevertheless reference is often made to them. Other valuable works published in the 1950s include Cryptogamic Botany. written by Gilbert Morgan Smith (1885-1959), the algal volume was published in 1955. In the following year (1956), Die Gattungen der Rhodophyceen. by Johan Harald Kylin) was published posthumously. Other phycologists who contributed massively to the knowledge of algae include: Elmer Yale Dawson (1918-1966) who published over 60 papers on the algae of the North American Pacific seas.

The number of books published in the mid to late 1800s shows how interest in the natural world developed. Books on algae were written by: Isabella Gifford (1853) *The Marine Botanist...*, some of her specimens are in the Ulster Museum; D. Landsborough (c.1779-1854) *A Popular History of British Seaweeds,...* third edition published in 1857; Louisa Lane Clarke (c.1812-1883) *The Common Seaweeds of the British Coast and Channel Islands;...* in 1865; S.O.Gray (1828-1902) British Sea-weeds:,... published 1867 and W.H.Grattann *British Marine Algae:...*published about 1874. These books were for the common people.

In 1902 Edward Arthur Lionel Batters (1860-1907) published *"A catalogue of the British Marine algae."* (Batters, 1902). In this he detailed records of algae found on the shores of the British Isles with the localities. This was the start of a new approach, the bringing together of records, detailed keys, checklists and mapping schemes.

The process accelerated in the 20th century. Lilly Newton (née Batten) (1893-1981) Professor in Botany at the University College of Wales, Aberystwyth and Professor Emeritus in 1931 wrote: *A Handbook of the British Seaweeds*. This was the first, and for quite a time, the only book for identification of seaweeds in the British Isles using a botanical key. In 1962 Eifion Jones published: A key to the genera of the British seaweeds. This small booklet provided a valuable source bridging the period before the valuable series *Seaweeds of the British Isles* was produced by the British Museum (Natural History) or The Natural History Museum.

Research advanced so quickly that the need for an up-to-date checklist became apparent. Mary Parke (1902-1981), who was a founder member of the British Phycological Society, produced a preliminary checklist of British marine algae in 1953, corrections and additions of this were published in 1956, 1957 and 1959. In 1964 M. Parke and Peter Stanley Dixon (1929-1993) published a revised check-list, a second revision of this was produced in 1968 and a third revision in 1976. Distribution was added to the checklist in 1986 with G.R.South and I. Tittley's A Checklist and Distributional Index of the Benthic Marine Algae of the North Atlantic Ocean. In 2003 A

Check-list and Atlas of the Seaweeds of Britain and Ireland was published by Gavin Hardy and Michael Guiry with a revised edition in 2006. This shows how rapidly knowledge of algae, at least in the British Isles, advanced. First efforts had been made by interested biologists and people capable of identifying the algae, this required books using the botanical names. Botanical keys to identify the plants then developed, followed by checklists. As more information was brought to light by interested workers, some volunteers, the checklists were improved and eventually a mapping scheme brought together all this information. The same pattern of knowledge developed with birds, mammals and flowering plants, though to a different time-scale and knowledge in other parts of the world has developed to this degree.

Numbers and Checklists

As records were collected the need to draw all the information together advanced. Checklists and annotated checklists were produced and updated so the actual numbers of different species became more precise. At first this was quite local. Threlkeld, in 1726, produced the first attempt at an enumeration of Irish Algae and in 1802 William Tighe published his "Marine plants observed at the County of Wexford," it included 58 marine and 2 freshwater species. In 1804 Wade published Plantae Rariores in Hibernia Inventae, in which 51 species of marine and 4 species of freshwater algae were enumerated. In the north of Ireland John Templeton and William Thompson were at work publishing on the algae of Ireland. In 1836 Mackay published his Flora Hibernica including 296 species. Adams, in his synopsis of 1908, listed a total of marine species reaching 843.

In more localised lists Adams (in 1907) listed the species of County Antrim noted that of the 747 species included in "Batter's List" he recorded 211 species from the Co. Antrim coast. In 1907 a list of marine aslgae from Lambay Island (County Dublin) was published by Batters. In 1960 A preliminary list of the marine algae of Galloway coast was published.

At the international level there are well over 3,000 species of alga in Australia.

Identification

As the study and identification of the different species became more extensive it became clear that identification was not at all easy. Harvey's 1846–51 Phycologia Britannica. along with his other publications makes no effort to provide "keys" to help in the identification. In 1931 Newton's Handbook which gave the first key to assist in the identification of algae of the British Isles, in the same year Knight and Park gave a key in their "Manx Algae." Eifion Jones in 1962, wrote a key to the genera of British seaweeds. Others soon followed: Dickinson wrote one entitled British Seaweeds. and Adey and Adey (1973) gave keys to the identification of the Corallinaceae of the British Isles. Abott and Hollenberg, in 1976, published keys to the identification of algae of California.

Evolution of Classification in the Algae

Linnaeus's "sexual system" which he grouped plants according to the number of stamens and carepels in the flowers, although wholly artificial was advantageous in that a newly discovered plant could be fitted in amongst those already known. He divided the plant kingdom into 25 classes, one of which was the Cryptogamia — plants with "concealed reproductive organs". Linnaeus accepted 14 genera of algae of which only four, Conferva, Ulva, Fucus and Chara, contained organisms now regarded as algae. As a consequence of the great increase in the number of species the artificiality of the Linnaean system was appreciated so that during the 18th Century and early 19th Century considerable numbers of new genera were described. J.V.F. Lamouroux in 1813 was the first to separate the groups on the basis of colour, however this was not taken up by other botanists and it was Harvey, who in 1836, divided the algae into four major devisions solely on the basis of their pigmentation: Rhodospermae (red algae), Melanospermae (brown algae), Chlorospermae (green algae) and Diatomaceae.

In 1883 and 1897 Schmitz separated the Rhodophyceae into two main groups. The first contained the Bangiales and the second the Nemoniales, Cryptonemiales, Gigartinales and Rhodymeniales. The Rhodophyta are now arranged in the Orders: Porphyridiales, Goniotrichales, Erythropeltidales, Bangiales, Acrochaetiales, Colaconematales, Palmariales, Ahnfeltiales, Nemaliales, Gelidiales, Gracilariales, Bonnemaisoniales, Cryptonemiales, Hildenbrandiales, Corallinales, Gigartinales, Plocamiales, Rhodymeniales and Ceramiales. The Chlorophyta are arranged in the Orders: Chlorococcales, Microsporales, Chaetophorales, Phaeophilales, Ulvales, Prasiolales, Acrosiphoniales, Cladiphorales, Bryopsidales, Chlorocystidales, Klebsormidiales and Ulotrichales. The Heterokontophyta: Sphacelariales, Dictyotales, Ectocarpales, Ralfsiales, Utleriales, Sporochniales, Tilopteridales, Desmarestiales, Laminariales and the Fucales.

Chapter–2

The Study of Algae

INTRODUCTION

Algae (*sing*. alga) are a large and diverse group of simple, typically autotrophic organisms, ranging from unicellular to multicellular forms.

There are nearly 30,000 algae species. The largest and most complex marine forms are called seaweeds, with 10,000 species. They are photosynthetic, like plants, and "simple" because they lack the many distinct organs found in land plants. Though the prokaryotic *cyanobacteria* (commonly referred to as blue-green algae) were traditionally included as "algae" in older textbooks, many modern sources regard this as outdated and restrict the term *algae* to eukaryotic organisms.

All true algae therefore have a nucleus enclosed within a membrane and chloroplasts bound in one or more membranes. Algae constitute a paraphyletic and polyphyletic group, as they do not all descend from a common algal ancestor, although their chloroplasts seem to have a single origin.

Algae lack the various structures that characterize land plants, such as phyllids and rhizoids in nonvascular plants, or leaves, roots, and other organs that are found in tracheophytes. They are distinguished from protozoa in that they are photosynthetic. Many are photoautotrophic, although some groups contain members that are mixotrophic, deriving energy both from photosynthesis and uptake of organic carbon either by osmotrophy, myzotrophy, or phagotrophy. Some unicellular species rely entirely on external energy sources and have limited or no photosynthetic apparatus.

All algae have photosynthetic machinery ultimately derived from the cyanobacteria, and so produce oxygen as a by-product of photosynthesis, unlike other photosynthetic bacteria such as purple and green sulfur bacteria.

Algae are most prominent in bodies of water but are also common in terrestrial environments. However, terrestrial algae are usually rather inconspicuous and far more common in moist, tropical regions than dry ones, because algae lack vascular tissues and other adaptations to live on land. Algae are also found in other situations, such as on snow and on exposed rocks in symbiosis with a fungus as lichen.

The various sorts of algae play significant roles in aquatic ecology. Microscopic forms that live suspended in the water column (phytoplankton) provide the food base for most marine food chains. In very high densities (so-called algal blooms) these algae may discolor the water and outcompete, poison, or asphyxiate other life forms. Seaweeds grow mostly in shallow marine waters, however some have been recorded to a depth of 300 m ome are used as human food or harvested for useful substances such as agar, carrageenan, or fertilizer.

The lineage of algae according to Thomas Cavallier-Smith. The exact number and placement of endosymbiotic events is not yet clear, so this diagram can be taken only as a general guide Endosymbiotic events are noted by dotted lines.

The study of marine and freshwater algae is called phycology or algology.

CLASSIFICATION OF ALGAE

While *Cyanobacteria* have been traditionally included among the algae, referred to as the Cyanophytes or blue-green algae, recent works on algae usually exclude them due to large differences such as the lack of membrane-bound organelles, the presence of a single circular chromosome, the presence of peptidoglycan in the cell walls, and ribosomes different in size and content from eukaryotes Rather than in chloroplasts, they conduct photosynthesis on specialized infolded cytoplasmic membranes called thylakoid membranes. Therefore, they differ significantly from the algae despite occupying similar ecological niches.

By modern definitions algae are eukaryotes and conduct photosynthesis within membrane-bound organelles called chloroplasts. Chloroplasts contain circular DNA and are similar in structure to cyanobacteria, presumably representing reduced cyanobacterial endosymbionts. The exact nature of the chloroplasts is different among the different lines of algae, reflecting different endosymbiotic events. The table below lists the three major groups of algae and their lineage relationship is shown in the figure on the left. Note many of these groups contain some members that are no longer photosynthetic. Some retain plastids, but not chloroplasts, while others have lost them entirely.

Cyanobacterium

These algae have *primary* chloroplasts, i.e. the chloroplasts are surrounded by *two membranes* and probably developed through a single endosymbiotic event. The chloroplasts of red algae have chlorophylls *a* and *d* (often), and phycobilins, while those of the green alga have chloroplasts with chlorophyll *a* and *b*. Higher plants are pigmented similarly to green algae and probably developed from them, and thus Chlorophyta is a sister taxon to the plants; sometimes they are grouped as Viridiplantae.

Green Alga

These groups have green chloroplasts containing chlorophylls *a* and *b* Their chloroplasts are surrounded by *four and three membranes*, respectively, and were probably retained from an ingested green alga.

Chlorarachniophytes, which belong to the phylum Cercozoa, contain a small nucleomorph, which is a relict of the alga's nucleus.

Euglenids, which belong to the phylum Euglenozoa, live primarily in freshwater and have chloroplasts with only three membranes. It has been suggested that the endosymbiotic green algae were acquired through myzocytosis rather than phagocytosis.

Red Alga

These groups have chloroplasts containing chlorophylls *a* and *c*, and phycobilins. The latter chlorophyll type is not known from any prokaryotes or primary chloroplasts, but genetic similarities with the red algae suggest a relationship there.

In the first three of these groups (Chromista), the chloroplast has four membranes, retaining a nucleomorph in cryptomonads, and they likely share a common pigmented ancestor, although other evidence casts doubt on whether the Heterokonts, Haptophyta, and Cryptomonads are in fact more closely related to each other than other groups

The typical dinoflagellate chloroplast has three membranes, but there is considerable diversity in chloroplasts among the group, and it appears there were a number of endosymbiotic events here. The Apicomplexa, a group of closely related parasites, also have plastids called apicoplasts. Apicoplasts are not photosynthetic but appear to have a common origin with dinoflagellates chloroplasts

It was W.H. Harvey (1811-1866), who first divided the algae into four divisions based on their pigmentation. This is the first use of a biochemical criterion in plant systematics. Harvey's four divisions were: red algae (Rhodophyta), brown algae (Heteromontophyta), green algae (Chlorophyta) and Diatomaceae

FORMS OF ALGAE

Most of the simpler algae are unicellular flagellates or amoeboids, but colonial and non-motile forms have developed independently among several of the groups. Some of the more common organizational levels, more than one of which may occur in the life cycle of a species, are:

- *Colonial*: small, regular groups of motile cells.
- *Capsoid*: individual non-motile cells embedded in mucilage.
- *Coccoid*: individual non-motile cells with cell walls.
- *Palmelloid*: non-motile cells embedded in mucilage.
- *Filamentous*: a string of non-motile cells connected together, sometimes branching.
- *Parenchymatous*: cells forming a thallus with partial differentiation of tissues.

In three lines even higher levels of organization have been reached, with full tissue differentiation. These are the brown algae, —some of which may reach 50 m in length (kelps) —the red algae, and the green algae. The most complex forms are found among the green algae in a lineage that eventually led to the higher land plants. The point where these non-algal plants begin and algae stop is usually taken to be the presence of reproductive organs with protective cell layers, a characteristic not found in the other alga groups.

The first plants on earth evolved from shallow freshwater algae much like *Chara* some 400 million years ago. These probably had an isomorphic alternation of generations and were probably heterotrichous. Fossils of isolated land plant spores suggest land plants may have been around as long as 475 million years ago.

ALGAE AND SYMBIOSES

Some species of algae form symbiotic relationships with other organisms. In these symbioses, the algae supply photosynthates (organic substances) to the host organism providing protection to the algal cells. The host organism derives some or all of its energy requirements from the algae. Examples include:

- *lichens*: a fungus is the host, usually with a green alga or a cyanobacterium as its symbiont. Both fungal and algal species found in lichens are capable of living independently, although habitat requirements may be greatly different from those of the lichen pair.

- *corals*: algae known as zooxanthellae are symbionts with corals. Notable amongst these is the dinoflagellate *Symbiodinium*, found in many hard corals. The loss of *Symbiodinium*, or other zooxanthellae, from the host is known as coral bleaching.

- *sponges*: green algae live close to the surface of some sponges, for example, breadcrumb sponge (*Halichondria panicea*). The alga is thus protected from predators; the sponge is provided with oxygen and sugars which can account for 50 to 80% of sponge growth in some species.

LIFE-CYCLE

Rhodophyta, Chlorophyta and Heterokontophyta, the three main algal Phyla, have life-cycles which show tremendous variation with considerable complexity. In general there is an asexual phase where the seaweed's cells are diploid, a sexual phase where the cells are haploid followed by fusion of the male and female gametes. Asexual reproduction is advantageous in that it permits efficient population increases, but less variation is possible. Sexual reproduction allows more variation but is more costly because of the waste of gametes that fail to mate, among other things. Often there is no strict alternation between the sporophyte and gametophyte phases and also because there is often an asexual phase, which could include the fragmentation of the thallus.

Chapter–3

TYPES OF ALGAE

BROWN ALGAE

The Phaeophyceae or brown algae, (singular: alga) is a large group of mostly marine multicellular algae, including many seaweeds of colder Northern Hemisphere waters. They play an important role in marine environments both as food, and for the habitats they form. For instance *Macrocystis*, a member of the Laminariales or kelps, may reach 60 m in length, and forms prominent underwater forests. Another example is *Sargassum*, which creates unique habitats in the tropical waters of the Sargasso Sea. This is one of the few areas where a large biomass of brown algae may be found in tropical waters. Many brown algae such as members of the order Fucales are commonly found along rocky seashores. Some members of the division are used as food for humans. Worldwide there are about 1500-2000 brown seaweed species.

Brown algae belong to a very large group, the Heterokontophyta, a eukaryotic group of organisms distinguished most prominently by having chloroplasts surrounded by four membranes, suggesting an origin from a symbiotic relationship between a basal eukaryote and another eukaryotic organism. Most brown algae contain the pigment fucoxanthin, which is responsible for the distinctive greenish-brown color that gives them their name. Brown algae are unique among heterokonts in developing into multicellular forms with differentiated tissues, but they reproduce by means of flagellate spores, which closely resemble other heterokont cells. Genetic studies show their closest relatives to be the yellow-green algae. The plants are filamentous, macroscopic or microscopic some polyiphonous. Some form crusts, cushions or are hollow and others grow to form large leathery fronds. The cells of most browns are connected by pores. They bear a unilocular (one holed) sporangium.

Phaeophyta evolved from the phaeothamniophyceae between 150 & 200 million years ago Claims that earlier (Ediacaran) fossils are brown algae have since been dismissed. The linages of brown algae diverged in the following order, from oldest to youngest: Dictyotales; Sphacelariales; Cutleriales; Desmerestales; Ectocarpales; Laminarales; Fucales. Their occurrence as fossils is rare due to their generally soft-bodied habit, and scientists continue to debate the identification of some finds. Other algae groups, such as the red algae and green algae have a number of calcareous members, which are more likely to leave evidence in the fossil record than the soft bodies of the brown algae. Miocene fossils of a soft-bodied brown macro algae, *Julescrania*, have been found well-preserved in Monterey Formation diatomites, but few other certain fossils, particularly of older specimens are known in the fossil record.

This is a list of the orders in the class Phaeophyceae:

- Ascoseirales Petrov
- Choristocarpales
- Cutleriales Oltmanns
- Desmarestiales Setchell & Gardner
- Dictyotales Kjellman
- Ectocarpales Setchell & Gardner
- Fucales Kylin
- Ishigeales
- Laminariales Migula

- Ralfsiales Nakamura
- Scytothamnales A.F. Peters & M. N. Clayton
- Sphacelariales Oltmanns
- Sporochnales Sauvageau
- Syringodermatales E.C. Henry
- Tilopteridale Bessey

A few species, such as *Botrydium stoloniferum*, are placed *incertae sedis*, or of uncertain position, as to order in this classification scheme.

The life cycle shows great variability from one group to another. However the life cycle of *Laminaria* consists of the diploid generation, that is the large plant well know to most people. It produces sporangia from specialised microscopic structures, these divide meiotically (meiosis) before they are released. As they are haploid there are equal numbers of male and female spores. With the exception of the Fucales all brown algae have a life cycle which consists of an alternation between morphologically haploid and diploid plants referred to as Monomorphic. A dimorphic life cycle consists of an alteration between dissimilar haploid and diploid plants.

Brown algae have adapted to a wide variety of marine ecological niches including the tidal splash zone, rock pools, the whole intertidal zone and relatively deep near shore waters. They are an important constituent of brackish water ecosystems, and four species survive in fresh water Most phaeophyceae are intertidal or upper littoral and they they are predominantly cool and cold water organisms that benefit from nutrients in up welling cold water currents and inflows from land; *Sargassum* being a prominent exception to this generalisation. Brown algae growing in brackish waters are almost solely asexual.

GOLDEN ALGAE

The golden algae or chrysophytes are a large group of heterokont algae, found mostly in freshwater. Originally they were taken to include all such forms except the diatoms and multicellular brown algae, but since then they have been divided into several different groups based on pigmentation and cell structure. They are now usually restricted to a core group of closely related forms, distinguished primarily by the structure of the flagella in motile cells, also treated as an order Chromulinales. It is possible membership will be revised further as more species are studied in detail. They come in a variety of morphological types, originally treated as separate orders or families. Most members are unicellular flagellates, with either two visible flagella, as in *Ochromonas*, or sometimes one, as in *Chromulina*. The Chromulinales as first defined by Pascher in 1910 included only the latter type, with the former treated as the order Ochromonadales. However, structural studies have revealed that short second flagellum or at least a second basal body is always present, so this is no longer considered a valid distinction.

Most of these have no cell covering. Some have loricae or shells, such as *Dinobryon*, which is sessile and grows in branched colonies. Most forms with silicaceous scales are now considered a separate group, the synurids, but a few belong among the Chromulinales proper, such as *Paraphysomonas*. Some members are generally amoeboid, with long branching cell extensions, though they pass through flagellate stages as well. *Chrysamoeba* and *Rhizochrysis* are typical of these. There is also one species, *Myxochrysis paradoxa*, which has a complex life cycle involving a multinucleate plasmodial stage, similar to those found in slime moulds.

These were originally treated as the order Chrysamoebales. The superficially similar *Rhizochromulina* was once included here, but is now given its own order based on differences in the structure of the flagellate stage. Other members are non-motile. Cells may be naked and embedded in mucilage, such as *Chrysosaccus*, or coccoid and surrounded by a cell wall, as in *Chrysosphaera*. A few are filamentous or even parenchymatous in organization, such as *Phaeoplaca*. These were included in various older orders, most of the members of which are now included in separate groups.

Hydrurus and its allies, freshwater genera which form branched gelatinous filaments, are often placed in the separate order Hydrurales but may belong here.

Chrysophytes were once considered to be a specialized form of cyanobacteria containing the golden-yellow pigment, fucoxanthin. Because many of these organisms had a silica capsule, they have a relatively complete fossil record, allowing modern biologists to confirm that they are, in fact, not derived from cyanobacteria, but rather an ancestor that did not possess the capability to photosynthesize. Many of the chrysophyta precursor fossils entirely lacked any type of photosynthesis-capable pigment. Most biologists believe that the chrysophytes obtained their ability to photosynthesis from an endosymbiotic relationship with fucoxanthin-containing cyanobacteria.

GREEN ALGAE

The green algae (singular: green alga) are the large group of algae from which the embryophytes (higher plants) emerged. As such, they form a paraphyletic group, although the group including both green algae and embryophytes is monophyletic (and often just known as kingdom Plantae). The green algae include unicellular and colonial flagellates, usually but not always with two flagella per cell, as well as various colonial, coccoid, and filamentous forms. In the Charales, the closest relatives of higher plants, full differentiation of tissues occurs. There are about 6000 species of green algae. Many species live most of their lives as single cells, while other species form colonies or long filaments.

A few other organisms rely on green algae to conduct photosynthesis for them. The chloroplasts in euglenids and chlorarachniophytes were acquired from ingested green algae, and in the latter retain a vestigial nucleus (nucleomorph). Some species of green algae, particularly of genera *Trebouxia* or *Pseudotrebouxia* (Trebouxiophyceae), can be found in symbiotic associations with fungi to form lichens. In general the fungal species that partner in lichens cannot live on their own, while the algal species is often found living in nature without the fungus. Almost all forms have chloroplasts. These contain chlorophylls *a* and *b*, giving them a bright green colour (as well as the accessory pigments beta carotene and xanthophylls) and have stacked thylakoids. All green algae have mitochondria with flat cristae. When present, flagella are typically anchored by a cross-shaped system of microtubules, but these are absent among the higher plants and charophytes. Flagella are used to move the organism. Green algae usually have cell walls containing cellulose, and undergo open mitosis without centrioles.

The chloroplasts of green algae are bound by a double membrane, so presumably they were acquired by direct endosymbiosis of cyanobacteria. A number of cyanobacteria show similar pigmentation, but this appears to have arisen more than once, and the chloroplasts of green algae are no longer considered closely related to such forms. Instead, the green algae probably share a common origin with the red algae.

Classification

Green algae are often classified with their embryophyte descendants in the green plant clade Viridiplantae (or Chlorobionta). Viridiplantae, together with red algae and glaucophyte algae, form the supergroup Primoplantae, also known as Archaeplastida or Plantae *sensu lato*. Classification systems which have a kingdom of Protista may include green algae in the Protista or in the Plantae. Growth of the green seaweed, *Enteromorpha* on rock substratum at the ocean shore. Some green seaweeds, such as *Enteromorpha* and *Ulva*, are quick to utilize inorganic nutrients from land runoff, and thus can be indicators of nutrient pollution.

- Chlorophyta
- Chlorophyceae
- Ulvophyceae
- Trebouxiophyceae
- Chlorokybales

- Klebsormidiales
- Zygnematales
- Desmidiales
- Coleochaetales
- Charales (stoneworts)

The orders outside the Chlorophyta are often grouped as the division Charophyta, which is paraphyletic to higher plants, together comprising the Streptophyta. Sometimes the Charophyta is restricted to the Charales, and a division Gamophyta is introduced for the Zygnematales and Desmidiales. In older systems the Chlorophyta may be taken to include all the green algae, but taken as above they appear to form a monophyletic group.

One of the most basal green algae is the flagellate *Mesostigma*, although it is not yet clear whether it is sister to all other green algae, or whether it is one of the more basal members of the Streptophyta.

Reproduction

Green algae are eukaryotic organisms that follow a reproduction cycle called alternation of generations. Reproduction varies from fusion of identical cells (isogamy) to fertilization of a large non-motile cell by a smaller motile one (oogamy). However, these traits show some variation, most notably among the basal green algae, called prasinophytes.

Haploid algae cells (containing only one copy of their DNA) can fuse with other haploid cells to form diploid zygotes. When filamentous algae do this, they form bridges between cells, and leave empty cell walls behind that can be easily distinguished under the light microscope. This process is called *conjugation*. The species of *Ulva* are reproductively isomorphic, the diploid vegetative phase is the site of meiosis and releases haploid zoospores, which germinate and grow producing a haploid phase alternating with the vegetative phase.

KELP

Kelp are large seaweeds (algae), belonging to the brown algae and classified in the order Laminariales. There are about 30 different genera. Some species grow very long indeed, and form kelp forests. Despite their plant-like appearance, some scientists group them not with the terrestrial plants (kingdom Plantae), but instead place them either in kingdom Protista or in kingdom Chromista.

Kelp grows in underwater "forests" (kelp forests) in clear, shallow oceans. It requires nutrient-rich water below about 20°C (68°F). It is known for its high growth rate—the genus *Macrocystis* and *Nereocystis luetkeana* grow as fast as half a metre a day, ultimately reaching 30 to 80 m. Through the 19th century, the word "kelp" was closely associated with seaweeds that could be burned to obtain soda ash (primarily sodium carbonate). The seaweeds used included species from both the orders Laminariales and Fucales. The word "kelp" was also used directly to refer to these processed ashes.

In most kelp, the thallus (or body) consists of flat or leaf-like structures known as blades. Blades originate from elongated stem-like structures, the stipes. The holdfast, a root-like structure, anchors the kelp to the substrate of the ocean. Gas-filled bladders (pneumatocysts) form at the base of blades of American species, such as *Nereocystis lueteana* (Mert. & Post & Rupr.) and keep the kelp blades close to the surface, holding up the leaves by the gas they contain.

Growth occurs at the base of the meristem, where the blades and stipe meet. Growth may be limited by grazing. Sea urchins, for example, can reduce entire areas to urchin barrens. The kelp life cycle involves a diploid sporophyte and haploid gametophyte stage. The haploid phase begins when the mature organism releases many spores, which then germinate to become male or female gametophytes. Sexual reproduction then results in the beginning of the diploid sporophyte stage which will develop into a mature plant.

Bongo kelp ash is rich in iodine and alkali. In great amount, kelp ash can be used in soap and glass production. Until the Leblanc process was commercialized in the early 1800s, burning of kelp in Scotland was one of the principal industrial sources of soda ash (predominantly sodium carbonate). Alginate, a kelp-derived carbohydrate, is used to thicken products such as ice cream, jelly, salad dressing, and toothpaste, as well as an ingredient in exotic dog food and in manufactured goods. Giant kelp can be harvested fairly easily because of its surface canopy and growth habit of staying in deeper water. Kelp is also used frequently in seaweed fertiliser, especially in the Channel Islands, where it is known as *vraic*.

Kombu (*Laminaria japonica* and others), several Pacific species of kelp, is a very important ingredient in Japanese cuisine. Kombu is used to flavor broths and stews (especially *dashi*), as a savory garnish (*tororo konbu*) for rice and other dishes, as a vegetable, and a primary ingredient in popular snacks (such as *tsukudani*). Transparent sheets of kelp (*oboro konbu*) are used as an edible decorative wrapping for rice and other foods. Kombu can be used to soften beans during cooking, and to help convert indigestible sugars and thus reduce flatulence. Because of its high concentration of iodine, brown kelp (Laminaria) has been used to treat goiter, an enlargement of the thyroid gland caused by a lack of iodine, since medieval times

Kelp in History and Culture

During the Highland Clearances, many Scottish Highlanders were moved off their crofts, and went to industries such as fishing and kelping (producing soda ash from the ashes of kelp). At least until the 1820s, when there were steep falls in the price of kelp, landlords wanted to create pools of cheap or virtually free labour, supplied by families subsisting in new crofting townships. Kelp collection and processing was a very profitable way of using this labour, and landlords petitioned successfully for legislation designed to stop emigration. But the economic collapse of the kelp industry in northern Scotland led to further emigration, especially to North America.

Natives of the Falkland Islands are sometimes nicknamed "Kelpers" The name is primarily applied by outsiders rather than the natives themselves.

CORALLINE ALGAE

Coralline algae are red algae in the Family Corallinaceae of the order Corallinales. They are characterized by a thallus that is hard because of calcareous deposits contained within the cell walls. The colors of these algae are most typically pink, or some other shade of red, and some species can be purple, yellow, blue, white or gray-green.

Unattached specimens (maerl, rhodoliths) may form relatively smooth compact balls to warty or fruticose thalli. Many are typically encrusting and rock-like, found in tropical marine waters all over the world. Coralline algae play an important role in the ecology of coral reefs. Sea urchins, parrot fish, limpets (molluscs) and chitons molluscs feed on coralline algae.

A close look at almost any intertidal rocky shore or coral reef will reveal an abundance of pink to pinkish-grey patches, splashed as though by a mad painter over rock surfaces. These patches of pink paint are actually living algae: crustose coralline red algae. The red algae belong to the division Rhodophyta, within which the coralline algae form a distinct, exclusively marine order, the Corallinales.

Coralline algae are widespread in all of the world's oceans, where they often cover close to 100% of rocky substrata. Many are epiphytic (grow on other algae or marine angiosperms), or epizoic (grow on animals), and some are even parasitic on other corallines. Despite their ubiquity, the coralline algae are poorly known by ecologists, and even by specialist phycologists (people who study algae). For example, a recent book on the seaweeds of Hawai'i does not include any crustose coralline algae, even though corallines are quite well studied there and dominate many marine areas.

Traditionally, corallines have been divided into two groups, although this division does not constitute a taxonomic grouping:

— the geniculate (articulated) corallines;

— the non-geniculate (non-articulated) corallines.

Geniculate corallines are branching, tree-like plants which are attached to the substratum by crustose or calcified, root-like holdfasts. The plants are made flexible by having non-calcified sections (genicula) separating longer calcified sections intergenicula). Nongeniculate corallines range from a few micrometres to several centimetres thick crusts. They are often very slow growing, and may occur on rock, coral skeletons, shells, other algae or seagrasses. Crusts may be thin and leafy to thick and strongly adherent. Some are parasitic or partly endophytic on other corallines. Many coralline crusts produce knobby protuberances ranging from a millimetre to several centimetres high. Some are free-living as rhodoliths (rounded, free-living specimens). There are over 1600 described species of nongeniculate coralline algae.

The first coralline alga recognized as a living organism was probably Corallina in the 1st century AD. In 1837 Philippi recognized that coralline algae were not animals and he proposed the two generic names Lithophyllum and Lithothamnion as Lithothamnium For many years they were included in the order Cryptonemiales as the family Coallinaceae until in 1986 they were raised to the order Corallinales.

Corallines in Community Ecology

Many corallines produce chemicals which promote the settlement of the larvae of certain herbivorous invertebrates, particularly abalone. This is adaptive for the corallines as the herbivores then remove epiphytes which might otherwise smother the crusts and pre-empt available light. This is also important for abalone aquaculture, as corallines appear to enhance larval metamorphosis and the survival of larvae through the critical settlement period. It also has significance at the community level, as the presence of herbivores associated with corallines can generate patchiness in the survival of young stages of dominant seaweeds. This has been seen this in eastern Canada, and it is suspected the same phenomenon occurs on

Indo-Pacific coral reefs, yet nothing is known about the herbivore enhancement role of Indo-Pacific corallines, or whether this phenomenon is important in coral reef communities.

Some corallines slough off a surface layer of epithallial cells, which in a few cases may be an anti-fouling mechanism which serves the same function as enhancing herbivore recruitment. This also affects the community, as many algae recruit on the surface of a sloughing coralline, and are then lost with the surface layer of cells. This can also generate patchiness within the community. The common Indo-Pacific corallines Neogoniolithon fosliei and Sporolithon ptychoides slough epithallial cells in continuous sheets which often lie on the surface of the plants looking so much like wet tissue paper.

Not all sloughing serves an anti-fouling function. Epithallial shedding in most corallines is probably simply a means of getting rid of damaged cells whose metabolic function has become impaired. Morton and his students studied sloughing in the South African intertidal coralline algae, Spongites yendoi, a species which sloughs up to 50% of its thickness twice a year. This deep-layer sloughing, which is energetically costly, does not have any effect on seaweed recruitment when herbivores are removed. The surface of these plants is usually kept clean by herbivores, particularly the pear limpet, Patella cochlear. Sloughing in this case is probably a means of getting rid of old reproductive structures and grazer-damaged surface cells, and reducing the likelihood of surface penetration by burrowing organisms.

Some coralline algae develop into thick crusts which provide microhabitat for many invertebrates. For example, off eastern Canada, Morton found that juvenile sea urchins, chitons, and limpets suffer nearly 100% mortality due to fish predation unless they are protected by knobby and under-cut coralline algae. This is probably an important factor affecting the distribution and grazing effects of herbivores within marine communities. Nothing is known about the microhabitat role of Indo-Pacific corallines. However, the most common species in the region, Hydrolithon onkodes, often forms an intimate relationship with the chiton Cryptoplax larvaeformis. The chiton lives in burrows that it makes in H. onkodes plants, and comes out at night to graze on the surface of the coralline. This combination of grazing and burrowing results in a peculiar growth form (called "castles") in H. onkodes in which the coralline produces nearly vertical, irregularly curved lamellae.

Non-geniculate corallines are of particular significance in the ecology of coral reefs, where they provide calcareous material to the structure of the reef, help cement the reef together, and are important sources of primary production. Coralline algae are especially important in reef construction, as they lay down calcium carbonate as calcite. Although they contribute considerable bulk to the calcium carbonate structure of coral reefs, their more important rôle in most areas of the reef, is in acting as the cement which binds the reef materials together into a solid and sturdy structure.

An area where corallines are particularly important in constructing reef framework is in the algal ridge that characterizes surf-pounded reefs in both the Atlantic and Indo-Pacfic regions. Algal ridges are carbonate frameworks that are constructed mainly by nongeniculate coralline algae (after Adey 1978). They require high and persistent wave action to form, so are best developed on the windward reefs in areas where there is little or no seasonal change in wind direction. Algal ridges are one of the main reef structures that prevent oceanic waves from striking adjacent coastlines, and they thus help to prevent coastal erosion.

Economic Importance

Despite their hard, calcified nature, coralline algae have a number of economic uses. One use dates back to the 18th century, and involves the collection of unattached corallines (maërl) for use as soil conditioners. This is particularly significant in Britain and France, where more 300,000 tonnes of Phymatolithon calcareum (Pallas) Adey & McKinnin and Lithothamnion corallioides are dredged annually. Several thousand kilometres of maërl beds, composed of as-yet undetermined species belonging to the genera Lithothamnion and Lithophyllum, exist off the coast of Brazil, and have been subjected to a low level of commercial exploitation. Maërl is also used as a food additive for cattle and pigs, as well as in the filtration of acidic drinking water.

Corallines are also used in medicine, where the earliest use involved the preparation of a vermifuge from ground geniculate corallines of the genera Corallina and Jania. This use stopped towards the end of the 18th century. Modern medical science has found a more high-tech use for corallines in the preparation of dental bone implants. Apparently, the cell fusions provide an ideal matrix for the regeneration of bone tissue.

Since coralline algae contain calcium carbonate, they fossilize fairly well. They are useful as stratigraphic markers of particular significance in petroleum geology. Coralline rock has also been used as building stones, with the best examples being in Vienna, Austria. As a colourful component of live rock sold in the marine aquarium trade, coralline algae are desired in home aquariums for their aesthetic qualities.

RED ALGAE

The red algae (Rhodophyta, from Greek: *(rhodon)* = rose *(phyton)* = plant, thus red plant) are one of the oldest groups of eukaryotic algae and also one of the largest, with about 5,000–6,000 species of mostly multicellular, marine algae, including many notable seaweeds. Other references indicate 10,000 species.

The red algae form a distinct group characterized by the following attributes: eukaryotic cells without flagella and centrioles, using floridean starch as food reserve, with phycobiliproteins as accessory pigments (giving them their red color), and with chloroplasts lacking external endoplasmic reticulum and containing unstacked thylakoids. Most red algae are also multicellular, macroscopic, marine, and have sexual reproduction.

Many of the coralline algae, which secrete calcium carbonate and play a major role in building coral reefs, belong here. Red algae such as dulse *(Palmaria palmata)* and laver (nori/gim) are a traditional part of European and Asian cuisine and are used to make other products like agar, carrageenans and other food additives. The oldest fossil identified as a red alga is also the oldest fossil eukaryote that belongs to a specific modern taxon. *Bangiomorpha pubescens*, a multicellular fossil from arctic Canada, strongly resembles the modern red alga *Bangia* despite occurring in rocks.

Red algae are important builders of limestone reefs. The earliest such coralline algae, the solenopores, are known from the Cambrian Period. Other algae of different origins filled a similar role in the late Paleozoic, and in more recent reefs. There are also calcite crusts which have been interpreted as the remains of coralline red algae dating to the terminal Proterozoic. hallophytes resembling coralline red algae are known from the late Proterozoic Doushantuou formation

The red algae are considered to be one of the three groups forming the Archaeplastida, together with the glaucophytes and the green lineage or Viridiplantae. The archaeplastida originated from the primary endosymbiosis event between a eukaryote and a cyanobacterium ~1,600 million years ago creating the first chloroplast and the first photosynthetic eukaryote. The red algae are classified in the Archaeplastida, along with the glaucophytes and Viridiplantae (green algae and land plants).

Below are two valid published taxonomies of the red algae, although neither necessarily has to be used, as the taxonomy of the algae is still in a state of flux (with classification above the level of order having received little scientific attention for most of the 20th century) If one defines the kingdom Plantae to mean the Archaeplastida, the red algae will be part of that kingdom; but if Plantae are defined more narrowly, to be the Viridiplantae, then the red algae might be considered their own kingdom or part of the kingdom Protista. The two classification systems below place the red algae in the plant kingdom.

There are around 6,500 to 10,000 known species, nearly all of which are marine, with about 200 that only live in fresh water. However estimates of the number of real species vary by 100%.

Some examples of species and genera of red algae are:

- *Atractophora hypnoides*
- *Gelidiella calcicola*
- *Lemanea*, a freshwater genus
- *Palmaria palmata*, dulse
- *Schmitzia hiscockiana*
- *Chondrus crispus*, Irish moss
- *Mastocarpus stellatus*

The values of red algae reflect their lifestyles. The largest difference results from their photosynthetic pathway: algae which use HCO_3 as a carbon source have far more negative $d^{13}C$ values than those which only use CO_2 An additional difference of about 1.71% separates groups which are intertidal from those below the lowest tide line, which are never exposed to atmospheric carbon. The latter group use the more negative CO_2 which is dissolved in sea water, whereas those with access to atmospheric carbon reflect the more positive signature of this reserve.

Red algae have a double cell wall The outer layer are usually composed of "pectic substances", from which agar can be manufactured The internal wall is mostly cellulose.

Pit connections and pit plugs are unique and distinctive features of red algae that form during the process of cytokinesis following mitosis. In red algae, cytokinesis is incomplete. Typically, a small pore is left in the middle of the newly formed partition. The pit connection is formed where the daughter cells remain in contact. Shortly after the pit connection is formed cytoplasmic continuity is blocked by the generation of a pit plug, which is deposited in the wall gap that connects the cells. Connections between cells having a common parent cell are called a primary pit connections. Because apical growth is the norm in red algae, most cells have two primary pit connections, one to each adjacent cell. Connections that exist between cells not sharing a common parent cells are labeled secondary pit connections. These connections are formed when an unequal cell division produced a nucleated daughter cell that then fuses to an adjacent cell. Patterns of secondary pit connections can be seen in the order Ceramiales.

After a pit connection is formed, tubular membranes appear. A granular protein, called the plug core, then forms around the membranes. The tubular membranes eventually disappear. While some orders of red algae simply have a plug core, others have an associated membrane at each side of the protein mass, called cap membranes. The pit plug continues to exist between the cells until one of the cells dies. When this happens, the living cell produce a layer of wall material that seals off the plug. It is thought that the pit connections function as structural reinforcement, and as an avenue for cell to cell communication and/or symplastic transport in red algae. While the presence of the cap membrane could inhibit this transport between cells, it has been hypothesized that the tubular plug cores serve as a means of transport. The reproductive cycle of red algae may be triggered by factors such as day length.

Red algae lack motile sperm. Hence they rely on water currents to transport their gametes to the female organs – although their sperm are capable "gliding" to a carpogonium's trichogyne. The trichogyne will continue to grow until it encounters a spermatium; once it has been fertilised, the cell wall at its base progressively thickens, separating it from the rest of the carpogonium at its base. Upone their collision, the walls of the spermatium and carpogonium dissolve. The male nucleus divides and moves into the carpogonium; one half of the nucleus merges with the carpogonium's nucleus.

The polyamine spermine is produced which triggers carpospore production. Spermatangia may have long delicate appendages, which increase their chances of "hooking up". They display alternation of phases; as well as a gametophyte phase, many have two sporophyte phases, the carposporophyte producing carpospores, which germinate into a tetrasporophyte – this produces spore tetrads, which dissociate and germinate into gametophytes. The gametophyte is typically (but not always) identical to the tetresporophyte.

Carpospores may also germinate directly into thalloid gemetophytes, or the carposporophytes may produce a tetraspore without going through a (free living) tetrasporophyte phase Tetrasporangia may be arranged in a row (Zonate), in a cross (cruciate), or in a tetrad

The carposporophyte may be enclosed within the gametophyte, which may cover it with branches to form a cystocarp. A couple of case studies may be helpful to understand some of the life histories algae may display.

In a simple case, such as *Rhodochorton investiens*;

In the Carposporophyte: a spermatium merges with a trichogyne (a long hair on the female sexual organ), which then divides to form carposporangia – which produce carpospores. Carpospores germinate into gametophytes, which produce sporophytes. Both of these are very similar; they produce monospores from monosporangia "just below a cross wall in a filament" and their spores are "liberated through apex of sporangial cell.

The spores of a sporophyte produce either tetrasporophytes. Monospores produced by this phase germinate immediately, with no resting phase, to form an identical copy of parent. Tetrasporophytes may also produce a carpospore, which germinates to form another tetrasporophyte.

The gametophyte may replicate using monospores, but produces sperm in spermatangia, and "eggs"(?) in carpogonium.

A rather different example is *Porphyra gardneri*:

In its diploid phase, a carpospore can germinate to form a filamentous "conchocelis stage", which can also self-replicate using monospores. The conchocelis stage eventually produces conchosporangia. The resulting conchospore germinates to form a tiny prothallus with rhizoids, which develops to a cm-scale leafy thallus. This too can reproduce via monospores, which are produced inside the thallus itself. They can also reproduce via spermatia, produced internally, which are released to meet a prospective carpogonium in its conceptacle.

Chapter–4

ALGACULTURE

INTRODUCTION

Algaculture is a form of aquaculture involving the farming of species of algae. The majority of algae that are intentionally cultivated fall into the category of microalgae (also referred to as phytoplankton, microphytes, or planktonic algae). Macroalgae, commonly known as seaweed, also have many commercial and industrial uses, but due to their size and the specific requirements of the environment in which they need to grow, they do not lend themselves as readily to cultivation.

Some of the commercial and industrial purposes of algae cultivation are for production of bioplastics, dyes and colorants, feedstock, pharmaceuticals, pollution control, algae fuel and for possible future food sources.

ALGAE AS A FOOD

One example is the wrapper on a sushi roll. Other species are edible as well, such as Spirulina and dulse (*Palmaria palmata*). Dulse is a red species sold particularly in Ireland and Atlantic Canada. It is eaten raw, fresh, dried, or cooked like spinach. Spirulina is a blue-green microalgae with a long history as a food source in East Africa and pre-colonial Mexico. As it is high in protein and other nutrients it is currently used as a food supplement and as a treatment for malnutrition.

Chlorella, another popular microalgae, has similar nutrition and is an ingredient of "Chlorella Growth Factor," a nutritional supplement which makes the unsubstantiated claim that it can increase growth in animals and children.]. As a nutritional supplement it is touted as a method of reducing mercury levels, supposedly by chelation of the mercury to the cell wall of the organism. However, what little scientific study there is on this topic, appears to contradict this claim. Chlorella is very popular in Japan and is currently one of the most prescribed supplements in that country. Chlorella, particularly a transgenic strain which carries an extra mercury reductase gene, has been studied as an agent for the environmental remediation due to its ability to reduce Hg2+ to the less toxic elemental mercury.

Purple laver (*Porphyra*) is also collected and used in a variety of ways. In Wales, for example "laverbread" is a traditional food, and in Ireland it is collected and made into a jelly by stewing or boiling. Preparation also can involve frying or converting to a pinkish jelly by heating the fronds with a little water and beating with a fork. It is also harvested along western coast of North America, from California to British Columbia and by Native Hawaiians and the Maori of New Zealand.

Irish moss (*Chondrus crispus*), often confused with *Mastocarpus stellatus*, is the source of carrageenan for the stiffening of instant puddings, sauces, and dairy products such as ice cream. Irish moss is also used by brewers as a fining agent; the addition of Irish moss to the wort 15 minutes before the end of the boil produces a clearer beer.

Sea lettuce (*Ulva lactuca*), is used in Scotland where it is added to soups and salads. Dabberlocks or badderlocks (*Alaria esculenta*) is eaten either fresh or cooked in Greenland, Iceland, Scotland and Ireland.

FERTILIZER AND AGAR

For centuries seaweed has been used as fertilizer. It is also an excellent source of potassium for manufacture of potash and potassium nitrate. There are commercial uses of algae, such as agar.

MONOCULTURE

Often it is desired to grow just one species of algae in each growing vessel. With mixed cultures, one species tends to dominate over time and if a non dominant species is believed to have particular nutritive value for some larval animal, it is necessary to obtain pure cultures in order to cultivate this species. Individual species cultures are also needed for research purposes.

A common method of obtaining pure cultures is serial dilution. A wild sample or a contaminated lab sample of algae containing the desired algae is diluted with filtered water and small aliquots are introduced into a large number of small growing containers. The dilution is done following a microscopic examination of the source culture to a degree that leads one to expect on average there will be a few of the growing containers with only one cell of the desired species.

While algae is often grown in monocultures using microbiological techniques to purify the desired strain, another approach has been used very successfully to produce algae feed for the cultivation of a variety of mollusks. Sea water is passed through filters to remove algae which are too large for the larvae being cultivated. Tanks in a green house, sometimes on a balcony in the mollusk house, are filled with the partially filtered water and nutrients are added. The tanks may be aerated and the water is used after only a day or two of growing. The resulting thin soup of mixed algae has been shown to be an excellent food source for larval mollusks. An advantage of this method of algaculture is the low maintenance requirements.

When cultivating algae, several factors must be considered, and different algae have different requirements. Essential factors include water, carbon dioxide, minerals and light (the basic reaction in water is carbon dioxide + light energy = glucose + oxygen . The water must be in a temperature range that will support the specific algal species being grown.

Light and Mixing

In most algal-cultivation systems, light only penetrates the top 3 inches (7.6 cm) to 4 inches (10 cm) of the water. This is because as the algae grow and multiply, they become so dense that they block light from reaching deeper into the pond or tank. Algae only need about 1/10th the amount of light they receive from direct sunlight. Direct sunlight is often too strong for algae.

In order to have ponds that are deeper than 4 inches algae growers use various methods to agitate the water in their ponds, thus circulating the algae so that it does not remain on the surface, which would cause it to be over-exposed. Paddle wheels can be used to circulate (stir) the water in a pond. Compressed air can be introduced into the bottom of a pond or tank to agitate the water, bringing algae from the lower levels up with it as it makes its way to the surface.

Apart from agitation, another means of supplying light to algae is to place the light *in* the system. Glow plates are sheets of plastic or glass that can be submerged into a tank, providing light directly to the algae at the right concentration.

The odor associated with bogs, swamps, or any stagnant waters taken over by algae, can be due to oxygen depletion in the water caused by the decay of deceased algal blooms. Under anoxic conditions, the bacteria inhabiting algae cultures break down the organic material and produce hydrogen sulfide and ammonia which causes the odor. This condition, called hypoxia, often results in the death of all aquatic animals. In a system where algae is intentionally cultivated, maintained, and harvested, neither eutrophication nor aquatic hypoxia are likely to occur.

Some algae also produce odorous chemicals, particularly certain blue-green algae (cyanobacteria) such as *Anabaena*. The most well-known of these odor-causing chemicals are MIB (2-methylisoborneol) and Geosmin. They give a sort of musty or earthy odor that can be quite strong if an algae bloom is present. Subsequent death of the cyanobacteria will also release

MIB, etc. that is trapped in the cells. These chemicals can be smelled at very low levels, in the ppb range, and are responsible for many "taste and odor" issues in drinking water treatment and distribution. There are many good references on taste and odor, MIB, etc. in regards to cyanobacteria, but one example is "A Guide to Geosmin and MIB-producing Cyanobacteria in the United States", Izaguirre and Taylor, Water Science Technology2004, 49(9):19-24. Cyanobacteria can also produce chemical toxins that have been a problem in drinking water in some cases.

Nutrients must be controlled so algae will not be "starved" and nutrients will not be wasted. Algae can be mainly cultured in open-ponds (such as raceway-type ponds and lakes) and photobioreactors. Raceway ponds may be less expensive.

Raceway-type ponds and lakes are open to the elements and so sometimes called "open-pond" systems. They are much more vulnerable to contamination by other microorganisms, such as invasive algal species or bacteria. Because of these factors, the number of species successfully cultivated in an "open-pond" system for a specific purpose (such as for food, for the production of oil, or for pigments) are relatively limited. In open systems one does not have control over water temperature and lighting conditions. The growing season is largely dependent on location and, aside from tropical areas, is limited to the warmer months.

A major benefit to this type of system are that it is one of the cheaper ones to construct, in the very least only a trench or pond needs to be dug. It can also have some of the largest production capacities relative to other systems of comparable size and cost. This type of culture can be viable when the particular algae in question requires (or is able to survive) some sort of extreme condition that other algae can not survive. For instance, Spirulina sp. can grow in water with a high concentration of sodium bicarbonate and Dunaliela salina will grow in extremely salty water. Open culture can also work if there is a simple inexpensive system of selecting out the desired algae for use and to inoculate new ponds with a high starting concentration of the desired algae. Some chain diatoms fall into this category as they can be filtered from a stream of water flowing through an outflow pipe. A "pillow case" of a fine mesh cloth is tied over the outflow pipe and most algae flow right through. The chain diatoms are held in the bag and used to feed shrimp larvae (in Eastern hatcheries) and to inoculate new tanks or ponds.

Algae can also be grown in a photobioreactor (PBR). A PBR is a bioreactor which incorporates some type of light source. Virtually any translucent container could be called a PBR, however the term is more commonly used to define a closed system, as opposed to an open tank or pond.

A variation on the basic "open-pond" system is to enclose it with a transparent or translucent barrier, to cover a pond or pool with a greenhouse. A pond covered with a greenhouse could be considered a PBR. While this usually results in a smaller system, for economic reasons, it does take care of many of the problems associated with an open system. It allows more species to be grown, it allows the species that are being grown to stay dominant, and it extends the growing season, only slightly if unheated, and if heated it can produce year round. Because PBR systems are closed, all essential nutrients must be introduced into the system to allow algae to grow and be cultivated.

A PBR can be operated in "batch mode", but it is also possible to introduce a continuous stream of sterilized water containing nutrients, air, and carbon dioxide. As the algae grows, excess culture overflows and is harvested. If sufficient care is not taken, continuous bioreactors often collapse very quickly, however once they are successfully started, they can continue operating for long periods. An advantage of this type of algae culture is that algae in the "log phase" is produced which is generally of higher nutrient content than old "senescent" algae. It can be shown that the maximum productivity for a bioreactor occurs when the "exchange rate" (time to exchange one volume of liquid) is equal to the "doubling time" (in mass or volume) of the algae. Different types of PBRs include:

- tanks provided with a light source
- polyethylene sleeves or bags
- glass or plastic tubes.

HARVESTING OF ALGAE

Algae can be harvested using microscreens, by centrifugation, by flocculation. and by froth flotation. Alum and ferric chloride are chemical flocculants used to harvest algae. A commercial product called "Chitosan", commonly used for water purification, can also be used as a flocculant but is far more expensive. The shells of crustaceans are ground into powder and processed to acquire chitin, a polysaccharide found in the shells, from which chitosan is derived via de-acetylation. Water that is more brackish, or saline requires additional chemical flocculant to induce flocculation. Harvesting by chemical flocculation is a method that is often too expensive for large operations. Interrupting the carbon dioxide supply to an algal system can cause algae to flocculate on its own, which is called "autoflocculation".

In froth flotation, the water and algae are aerated into a froth, with the algae then removed from the water. Ultrasound based methods of algae harvesting are currently under development, and other, additional methods are currently being developed.

OIL EXTRACTION

Algae oils have a variety of commercial and industrial uses, and are extracted through a wide variety of methods.

The simplest method is mechanical crushing. Since different strains of algae vary widely in their physical attributes, various press configurations (screw, expeller, piston, etc) work better for specific algae types. Often, mechanical crushing is used in conjunction with chemicals. Estimates of the cost to extract oil from microalgae vary, but are likely to be around $1.80/kg (compared to $0.50/kg for palm oil).

Chemical solvents: Algal oil can be extracted using chemicals. Benzene and ether have been used, oil can also be separated by hexane extraction, which is widely used in the food industry and is relatively inexpensive. The downside to using solvents for oil extraction are the dangers involved in working with the chemicals. Care must be taken to avoid exposure to vapors and direct contact with the skin, either of which can cause serious damage. Benzene is classified as a carcinogen. Chemical solvents also present the problem of being an explosion hazard.

Soxhlet extraction is an extraction method that uses chemical solvents. Oils from the algae are extracted through repeated washing, or percolation, with an organic solvent such as hexane or petroleum ether, under reflux in a special glassware.

Enzymatic extraction: Enzymatic extraction uses enzymes to degrade the cell walls with water acting as the solvent, this makes fractionation of the oil much easier. The costs of this extraction process are estimated to be much greater than hexane extraction. The enzymatic extraction can be supported by ultrasonication. The combination "sonoenzymatic treatment" causes faster extraction and higher oil yields.

Expression / Expeller press: When algae is dried it retains its oil content, which then can be "pressed" out with an oil press. Many commercial manufacturers of vegetable oil use a combination of mechanical pressing and chemical solvents in extracting oil.

Osmotic shock: Osmotic shock is a sudden reduction in osmotic pressure, this can cause cells in a solution to rupture. Osmotic shock is sometimes used to release cellular components, such as oil.

Supercritical fluid: In supercritical fluid/CO_2 extraction, CO_2 is liquefied under pressure and heated to the point that it has the properties of both a liquid and a gas, this liquified fluid then acts as the solvent in extracting the oil.

Ultrasonic-assisted extraction: Ultrasonic extraction, a branch of sonochemistry, can greatly accelerate extraction processes. Using an ultrasonic reactor, ultrasonic waves are used to create cavitation bubbles in a solvent material, when these bubbles collapse near the cell walls, it creates shock waves and liquid jets that causes those cells walls to break and release their contents into the solvent.

Other methods are still being developed, including ones to extract specific types of oils, such as those with a high production of long-chain highly unsaturated fatty acids.

COMMERCIAL AND INDUSTRIAL USES

Algae are cultivated to serve many commercial and industrial uses:

- Bioplastics
- Dyes and Colorants
- Feedstock
- Nutritional
- Pharmaceutical
- Pollution Control
- CO_2 sequestration
- Uranium/Plutonium sequestration
- Fertiliser Runoff reclamation
- Sewage treatment

There are many algae that are cultivated for their nutritional value, either for supplemental use, or as a food source.

Spirulina *(Arthrospira platensis)* is a blue-green algae (cyanobacteria) that is quite nutritious. This species thrives in open systems and commercial growers have found it well-suited to cultivation. One of the largest production sites for Spirulina is Lake Texcoco in central Mexico The plants themselves produce a variety of nutrients and high amounts of protein, and is often used commercially as a nutritional supplement. Extracts and oils from algae are also used as additives in various food products. The plants also produce

Omega-3 and Omega-6 fatty acids, which are commonly found in fish oils, and which have been shown to have positive medical benefits to humans.

Some of the carbon dioxide that is released into the atmosphere is from the burning of fossil fuels. With concerns over global warming, new methods for the thorough and efficient capture of CO_2 are being sought out. An alternative to carbon capture and storage, by attaching an algae pond, or photobioreactor, to any fuel burning plant, the carbon dioxide produced during combustion can be fed into the algae system. Additional nutrients can be sourced from sewage, thus turning two pollutants into resources for the production of biofuels, with a land requirement much smaller than other crop sources (terrestrial crops) and without competing with food production.

Large scale algaculture of the oceans, as well as similar culture of other plankton, is proposed as a future food source in order to support a growing world population. Currently, however, it has limited application, since e.g. technologies of the green revolution has resulted in that land-based food production so far has increased faster than the demand of the growing world population.

Chapter–5

Algae Fuel

INTRODUCTION

Algae fuel, also called algal fuel, *Oilgae algaeoleum* or third generation biofuel, is a biofuel from algae. Compared with second generation biofuels, algae are high-yield high-cost (30 times more energy per acre than terrestrial crops) feedstocks to produce biofuels. Since the whole organism uses sunlight to produce lipids, or oil, algae can produce more oil in an area the size of a two-car garage than an entire football field of soybeans are many factors to overcome for algae to be a wide-spread source of energy, several positive factors can already be considered. Algal fuels do not impact fresh water resources and can use ocean and wastewater. The cost of various algae species is typically between US$5-10 per kg dry weight with research actively looking to reduce capital and operating costs and make algae oil production commercially viable.

With the record oil price increases since 2003, competing demands between foods and other biofuel sources and the world food crisis, there is much interest in algaculture (farming algae) for making vegetable oil, biodiesel, bioethanol, biogasoline, biomethanol, biobutanol and other biofuels.

The production of biofuels to replace oil and natural gas is in active development, focusing on the use of cheap organic matter (usually cellulose, agricultural and sewage waste) in the efficient production of liquid and gas biofuels which yield high net energy gain. One advantage of many biofuels over most other fuel types is that they are biodegradable, and so relatively harmless to the environment if spilled.

The United States Department of Energy estimates that if algae fuel replaced all the petroleum fuel in the United States, it would require 15,000 square miles (40,000 square kilometers), which is a few thousand square miles larger than Maryland, or 1.3 Belgiums. This is less than 1/7th the area of corn harvested in the United States in 2000.

The Aquatic Species Programme launched in 1978. The U.S. research program, funded by the U.S. DoE, was tasked with investigating the use of algae for the production of energy. The programme initially focused efforts on the production of hydrogen, however, shifted primary research to studying oil production in 1982. From 1982 through its culmination, the majority of the program research was focused on the production of transportation fuels, notably biodiesel, from algae. In 1995, as part of the over-all efforts to lower budget demands, the DoE decided to end the program. Research stopped in 1996 and staff began compiling their research for publication. In July of 1998, the DoE published the report "A Look Back at the U.S. Department of Energy's Aquatic Species Program: Biodiesel from Algae".

In 2008, *Time Magazine* voted Isaac Berzin one of the world's most influentials for 2008 for his ability to turn a dream of an oil-free future into a reality through GreenFuel, founded in Boston in 2001.

Dry algae factor is the percentage of algae cells in relation with the media where is cultured, e.g. if the dry algae factor is 50%, one would need 2 kg of wet algae (algae in the media) to get 1 kg of algae cells.

Lipid factor is the percentage of vegoil in relation with the algae cells needed to get it, i.e. if the algae lipid factor is 40%, one would need 2.5 kg of algae cells to get 1 kg of oil.

Fuels

The vegoil algae produce can then be harvested and converted into biodiesel; the algae's carbohydrate content can be fermented into bioethanol.

BIODIESEL

Currently most research into efficient algal-oil production is being done in the private sector, but predictions from small scale production experiments bear out that using algae to produce biodiesel may be the only viable method by which to produce enough automotive fuel to replace current world diesel usage. Microalgae have much faster growth-rates than terrestrial crops. The per unit area yield of oil from algae is estimated to be from between 5,000 to 20,000 gallons per acre, per year (4.6 to 18.4 l/m^2 per year); this is 7 to 30 times greater than the next best crop, Chinese tallow (699 gallons).

Algae can also grow on marginal lands, such as in desert areas where the groundwater is saline. The difficulties in efficient biodiesel production from algae lie in finding an algal strain with a high lipid content and fast growth rate that isn't too difficult to harvest, and a cost-effective cultivation system (i.e., type of photobioreactor) that is best suited to that strain.

Another obstacle preventing widespread mass production of algae for biofuel production has been the equipment and structures needed to begin growing algae in large quantities. Diversified Energy Corporation have avoided this problem by taking a different approach, and growing the algae in thin walled polyethylene tubing called Algae Biotape, similar to conventional drip irrigation tubing, which can be incorporated into a normal agricultural environment.

Open-pond systems for the most part have been given up for the cultivation of algae with high-oil content. Many believe that a major flaw of the Aquatic Species Program was the decision to focus their efforts exclusively on open-ponds; this makes the entire effort dependent upon the hardiness of the strain chosen, requiring it to be unnecessarily resilient in order to withstand wide swings in temperature and pH, and competition from invasive algae and bacteria. Open systems using a monoculture are also vulnerable to viral infection. The energy that a high-oil strain invests into the production of oil is energy that is not invested into the production of proteins or carbohydrates, usually resulting in the species being less hardy, or having a slower growth rate. Algal species with a lower oil content, not having to divert their energies away from growth, have an easier time in the harsher conditions of an open system. Some open sewage ponds trial production has been done in Marlborough, New Zealand.

In a closed system (not exposed to open air) there is not the problem of contamination by other organisms blown in by the air. The problem for a closed system is finding a cheap source of sterile carbon dioxide (CO_2). Several experimenters have found the CO_2 from a smokestack works well for growing algae. To be economical, some experts think that algae farming for biofuels will have to be done next to power plants, where they can also help soak up the pollution.

A feasibility study using marine microalgae in a photobioreactor is being done by The International Research Consortium on Continental Margins at the International University Bremen.

Research into algae for the mass-production of oil is mainly focused on microalgae; organisms capable of photosynthesis that are less than 0.4 mm in diameter, including the diatoms and cyanobacteria; as opposed to macroalgae, e.g. seaweed. However, some research is being done into using seaweeds for biofuels, probably due to the high availability of this resource his preference towards microalgae is due largely to its less complex structure, fast growth rate, and high oil content (for some species). Some commercial interests into large scale algal-cultivation systems are looking to tie in to existing infrastructures, such as coal power plants or sewage treatment facilities. This approach not only provides the raw materials for the system, such as CO_2 and nutrients; but it changes those wastes into resources.

Aquaflow Bionomic Corporation of New Zealand announced that it has produced its first sample of homegrown bio-diesel fuel with algae sourced from local sewerage ponds.

The Department of Environmental Science at Ateneo de Manila University in the Philippines, is working on producing biofuel from algae, using a local species of algae. The NBB's

Feedstock Development programme is addressing production of algae on the horizon to expand available material for biodiesel in a sustainable manner.

Biobutanol

Butanol can be made from algae or diatoms using only a solar powered biorefinery. This fuel has an energy density similar to gasoline, and greater than that of either ethanol or methanol. In most gasoline engines, butanol can be used in place of gasoline with no modifications. In several tests, butanol consumption is similar to that of gasoline, and when blended with gasoline, provides better performance and corosion resistance than that of ethanol or E85. The green waste left over from the algae oil extraction can be used to produce butanol.

Biogasoline

Methane

Through the use of algaculture grown organisms and cultures, various polymeric materials can be broken down into methane.

STRAIGHT VEGETABLE OIL

The algal-oil feedstock that is used to produce biodiesel can also be used for fuel directly as "Straight Vegetable Oil", (SVO). The benefit of using the oil in this manner is that it doesn't require the additional energy needed for transesterification, (processing the oil with an alcohol and a catalyst to produce biodiesel). The drawback is that it does require modifications to a normal diesel engine. Transesterified biodiesel can be run in an unmodified modern diesel engine, provided the engine is designed to use ultra-low sulfur diesel, which, as of 2006, is the new diesel fuel standard in the United States.

HYDROCRACKING TO TRADITIONAL TRANSPORT FUELS

Vegetable oil can be used as feedstock for an oil refinery where methods like hydrocracking or hydrogenation can be used to transform the vegetable oil into standard fuels such as gasoline and diesel.

Organisations

Algal Biomass Organization (ABO) is formed by Boeing Commercial Airplanes, A2BE Carbon Capture Corporation , National Renewable Energy Labs, Institution of Oceanography, Benemann Associates, Mont Vista Capital and Montana State University.

Global air carriers Air New Zealand, Continental, Virgin Atlantic Airways, and biofuel technology developer UOP LLC, a Honeywell company, will be the first wave of aviation-related members, together with Boeing, to join Algal Biomass Organization.

Twenty-five airlines went bust or stopped operations in the first six months of 2008 and more could fold as fuel prices soar, aviation industry association IATA has warned. Algal jet fuel can be used as alternative: IATA recognizes that aircraft are long lived and will be using kerosene or kerosene-type fuels for many years. It supports research, development & deployment into alternative fuels that produce less GHG emissions over their life cycle and do not compete for land with fuel crops. IATA's goal is for its members to be using 10% alternative fuels by 2017.

Algae Cultivation

Algae grow rapidly and can have a high percentage of lipids, or oils. They can double their mass several times a day and produce at least 15 times more oil per acre than alternatives such as rapeseed, palms, soybeans, or jatropha. Moreover, algae-growing facilities can be built on coastal land unsuitable for conventional agriculture.

The hard part about algae production is growing the algae in a controlled way and harvesting it efficiently.

Most companies pursuing algae as a source of biofuels are pumping nutrient-laden water through plastic tubes (called "bioreactors") that are exposed to sunlight (and so called photobioreactors or PBR). Running a PBR is more difficult than a open pond, and more costly.

Wastewater

There is an option currently being deployed at the Woods Hole Oceanographic Institution and the Harbor Branch Oceanographic Institution and s using wastewater for breeding algae. The wastewater from domestic and industrial sources contain rich organic compounds, which accelerate the growth of algae.

Algaewheel, based in Indianapolis, Indiana, presented a proposal to build a facility in Cedar Lake, Indiana that uses algae to treat municipal wastewater and uses the sludge byproduct to produce biofuel.

Algal Strains

- Botryococcene
- *Chlorella*
- *Dunaliella tertiolecta*
- *Gracilaria*
- *Pleurochrysis carterae* (also called CCMP647)
- *Sargassum*, with 10 times the output volume of *Gracilaria*.

Nutrients

Nutrients like nitrogen (N), phosphorous (P), and potassium (K), are important for plant growth and are essential parts of fertilizer. Silica and iron may also be considered important marine nutrients as the lack of one can limit the growth of, or productivity in, an area.

Another possible nutrient source is waste water from the treatment of sewage, agricultural, or flood plain run-off, all currently major pollutants and health risks. However, this waste water cannot feed algae directly and must first be processed by bacteria, through anaerobic digestion. If waste water is not processed before it reaches the algae, it will contaminate the algae in the reactor, and at the very least, kill much of the desired algae strain. In biogas facilities, organic waste is often converted to a mixture of carbon dioxide, methane, and organic fertilizer. Organic fertilizer that comes out of digester is liquid, and nearly suitable for algae growth, but it must first be cleaned and sterilized.

One company, Green Star Products, announced their development of a micronutrient formula to increase the growth rate of algae. According to the company, its formula can increase the daily growth rate by 34% and can double the amount of algae produced in one growth cycle.

SEAWEED

Seaweed is a loose colloquial term encompassing macroscopic, multicellular, benthic marine algae. The term includes some members of the red, brown and green algae. A seaweed may belong to one of several groups of multicellular algae: the red algae, green algae, and brown algae. As these three groups are not thought to have a common multicellular ancestor, the seaweeds are a paraphyletic group. In addition, some tuft-forming bluegreen algae (Cyanobacteria) are sometimes considered as seaweeds—"seaweed" is a colloquial term and lacks a formal definition. Seaweeds' appearance somewhat resembles non-arboreal terrestrial plants.

- *thallus:* the algal body;
- *lamina:* a flattened structure that is somewhat leaf-like;
- *sorus:* spore cluster;
- on *Fucus*, air bladders: float-assist organ (on blade);
- on kelp, floats: float-assist organ (between lamina and stipe);
- *stipe:* a stem-like structure, may be absent;
- *holdfast:* specialized basal structure providing attachment to a surface, often a rock or another alga.
- *haptera:* finger-like extensions of holdfast anchoring to benthic substrate;

The stipe and blade are collectively known as fronds.

Ecology

The ecology of seaweeds is dominated by two specific environmental requirements. These are the presence of seawater (or at least brackish water) and the presence of light sufficient to drive photosynthesis. A very common requirement is also to have a firm point of attachment. As a result, seaweeds are most commonly found in the littoral zone and within that zone more frequently on rocky shores than on sand or shingle. The ecological niches utilised by seaweeds are wide ranging. At the highest level are those that inhabit the zone that is only wetted by the tops of sea spray, the deepest living are those that are attached to the seabed under several meters of water. In some parts of the world, the area colonized by littoral seaweeds can extend for several miles away from the shore. The limiting factor in such cases is the availability of sufficient sun-light to support photosynthesis. The deepest living seaweeds are the various kelps. In addition to the familiar seashore seaweeds, a number of species have adapted to a fully planktonic niche and are free-floating, often with the assistance of gas filled sacs. *Sargassum* is one of the better known examples of this type of seaweed. A number of species have adapted to the specialised environment of tidal rock pools. In this niche seaweeds are able to withstand rapidly changing temperature and salinity and even occasional drying.

Uses

Small plots being used to farm seaweed in Indonesia, with each square belonging to a different family. Seaweed has a variety of purposes, for which it is farmed, or foraged from the wild.

Packaged Seaweed

Seaweeds are extensively used as food by coastal people, particularly in East Asia, e.g. Japan, China, Korea, Taiwan, and Vietnam, but also in Indonesia, Peru, the Canadian Maritimes, Scandinavia, Ireland, Wales, Philippines, and Scotland, among other places. Tiwi, Albay residents discovered a new pansit or noodles made from seaweed, which has health benefits. It is rich in calcium and magnesium and the seaweed noodles can be cooked into pansit canton, pansit luglug, spaghetti or carbonara.

In Asia, Zicai (in China), gim (in Korea) and nori (in Japan) are sheets of dried *Porphyra* used in soups or to wrap sushi. *Chondrus crispus* (commonly known as Irish moss or carrageenan moss) is another red alga used in producing various food additives, along with Kappaphycus and various gigartinoid seaweeds. Porphyra is a red alga used in Wales to make laver. Laverbread, made from oats and the laver, is a popular dish in Wales.

Seaweeds are also harvested or cultivated for the extraction of alginate, agar and carrageenan, gelatinous substances collectively known as hydrocolloids or phycocolloids. Hydrocolloids have attained commercial significance, especially in food production as food additives. The food industry exploits the gelling, water-retention, emulsifying and other physical properties of these hydrocolloids. Agar is used in foods such as confectionery, meats and poultry products, desserts and beverages and moulded foods. Carrageenan is used in preparation of salad dressings and sauces, dietetic foods, and as a preservative in meat and fish products, dairy items and baked goods. Alginates enjoy many of the same uses as carrageenan, but are also used in production of industrial products such as paper coatings, adhesives, dyes, gels, explosives and in processes such as paper sizing, textile printing, hydro-mulching and drilling.

Medicine

In the biomedicine and pharmaceutical industries, alginates are used in wound dressings, and production of dental moulds and have a host of other applications. In microbiology research, agar is extensively used as culture medium. Carrageenans, alginates and agaroses (the latter are prepared from agar by purification), together with other lesser-known macroalgal polysaccharides, also have several important biological activities or applications in biomedicine. Seaweed is also a known source of iodine an element necessary for thyroid function with deficiencies leading to goitre. It has been asserted that seaweeds may have curative properties for tuberculosis, arthritis, colds and influenza, worm infestations and even tumors [*dubious – discuss*]. A number of research studies have been conducted to investigate these claims and other effects of seaweed on human health.

Chapter–6

ALGAE IN AQUARIUM

INTRODUCTION

Almost every aquarist will at some point face an algae outbreak in a tank, and as we have learned from this previous article on the site, algae control is all nutrient control. But one thing has to be clear: it is not the same when it comes to controlling algae in planted and non-planted tanks!

In a non-planted aquarium, nitrate and phosphate levels can easily go high, and result in an algae outbreak. Overfeeding and overstocking is the most common reason for water quality going bad like that. Weekly water changes are the best solution in reducing these nutrients. The ammonia (NH_3), ammonium (NH_4) and nitrite (NO_2) levels should be at 0 ppm. Nitrate (NO_3) levels should ideally be kept below 10 ppm (nitrate levels over 40 ppm can be dangerous for some fish and invertebrates). Phosphate (PO_4) levels should be kept below 0.5 ppm. Lights should be on no more then 10 hours per day. Note that most algae favours strong lights, so placing the aquarium away from the window is a good idea. Direct sunlight will almost certainly cause an algae outbreak. An algae eating army will help a lot in combating algae.

In a planted aquarium, the situation is more complex because there are not just fish to consider, but also live plants. It is true that plants will uptake the ammonia/nitrate/ phosphate and help keep the water chemistry in good quality.

But even so you can suddenly experience the worst algae outbreak in a planted tank, and you will ask "but how is this possible?!"

Let's start answering with the fact that plants and algae need the same nutrients to thrive. Imbalance between lights, CO_2, macro- and micro-nutrientes are the main causes that lead to algae growth, because when plants have all those elements balanced they are much more efficient at uptaking nutrients from water than the algae. Fast growing stem plants are very famous for keeping algae at bay for their ability to uptake nutrients in no time. But when, for example, just one of these elements is reduced, plants will slow down their metabolism and will leave nutrients available, sometimes even start leaking nutrients back into the water. Of course the plant will show symptoms like yellow leaves and pin holes in leaves, etc. It is not enough to add a few fast growing plants to fight algae, without providing the right nutrient balance for them to thrive. Only happy plants will lead to a balanced ecosystem. Since algae are nutrient scavengers and much simpler life forms than plants, they will take advantage in unbalanced systems. So, all we need to do is to make sure our plants have the right amount of lighting, CO_2, macro- and micro-nutrients. The fertilising regime depends on how strong the aquarium lights.

About 0.3-0.4 W/L (Watts per liter) of normal fluorescent lighting. CO_2 can be added via DIY Yeast Reactor, that is inexpensive. Fertilisers like Tropica Master Grow are very good for dosing trace elements (micronutrients) like Fe, Mg, Zn, B, S, Mn, Cu, Mo and K. Homemade fertilizer mix (Poor Mans Dosing Drops, PMDD) is a very good solution for dosing macronutrients (N-P-K) and others. It is made out of KNO3, KH2PO4 and NutriSI mikro. NOTE; Because Iron (Fe) is the most important trace element, the tank substrate should have fair amount of Iron. Liquid iron will, if overdosed, favour hair algae. It can be added through ground Iron-rich fertilisers and through substrates like Laterite and Flourite. Low light set-ups that are injected with a fair amount of CO_2 through DIY Yeast Reactor (1 bottle per 100 litres) should be fertilised at least once a week. It fertilise with PMDD and Tropica Master Grow. Stem plants, in this kind of set-up, will normally need to be pruned every few weeks. It is usually recommendable to perform 50% water change per week. This way we limit nutrient build-up, and after that the nutrients should be re-dosed. A good fertilising regime is to break the nutrients dosing every 2-3 days. When dosing nutrients only once a week, plants may have excess nutrients in the beginning and lack nutrients in the second part of the week, causing more variable, unbalanced conditions that favour algae.

We will find that many aquarists establish low light, low tech tanks and change water approximately every 3-4 months (low maintenance). To successfully run low tech planted tanks it is very important to have nutrient rich substrate. Some use ground peat, others clay or Laterite/Fluorite, and some mix together different types of substrates. The important thing is to cover nutrient rich substrate with approximately 4-5 cm of fine gravel, which will keep the water clean and nutrients away from algae. Liquid fertiliser should be added every month, half of the recommended dose and if plants show deficiencies, add the recommended measure. Low tech tanks are normally not injected with CO_2, so the filter outlet should not disturb the water surface too much (white water, splashes, etc). In low tech tanks, plants will grow slow and will not need much pruning.

HIGH LIGHT TANKS

have 0.7 W/L or higher of lighting, and therefore need equally higher CO_2 and nutrient levels to keep the correct balance for plants to grow fast enough and out-compete algae. A pressurized CO_2 system becomes the best option and one should be ready to invest in it to have an algae free aquarium with lush plants. You may need double or triple the amount of fertilizing as in low light tanks. In high light set-ups CO_2 should be monitored carefully and maintained at aroung 30 ppm. In both aquarium types (low and high light) the nitrate levels should be maintained at 10-15 ppm and phosphates at 0.5-1 ppm. To repeat one thing though, because it is 95% algae issue related; higher levels of CO_2 (30 ppm) seem to keep algae at bay. Of course, algae eaters like SAE, Otos and Amano shrimp are more than a welcome addition. Snails like Malaysian Trumpet will help aerate the gravel. Trimming plants will do good since it will encourage new plant growth and stop older leaves from decaying and polluting the water. And remember, limiting nutrients is the right thing to do in *non-planted tanks*, but very wrong in planted tanks as in question. In high light planted tanks one should do weekly 50% water change, and among other options use the Estimative Index method for dosing nutrient.

Note: There are many articles on the net preaching that reducing nitrates and phosphates will help keeping algae low! In non-planted tanks yes, but in planted tanks NO! Plants need more nutrients than algae to thrive, so do not reduce nutrients, dose them!

Now that we know how to prevent algae in general, let's ID the most common algae types in freshwater aquariums and discuss specific causes and combat techniques:

BROWN ALGAE

Brown Algae Are more likely to appear in low-light aquariums and new setups, where nitrogen (N) is low and phosphate (P) level is high, with excess silicate (SiO_2). It's been known that strong lights make this algae go away, but they might still be seen on lower, shadowed, plant leaves (*Bacopa australis*, *Cardamine lyrata*). It can also be found on aquarium glass, gravel and decoration. It can be easily removed manually, since it has a soft/slimy structure. Siamese Algae Eaters, Otos (this catfish relishes this type of algae) and Snails can easily keep this algae in low numbers. Healthy plants can prevent this algae from over taking the tank.

GREEN WATER ALGAE

This is the most common problem if the cloudy situation extends beyond 10-14 days. Note that "green water" (GW) is not always green in appearance! Since GW is the most common problem and the most difficult to solve the answer needs to reflect several options. The situation that causes GW is usually a combination of high nitrates, phosphates, and mixed in some ammonia/ammonium. Substrate disturbance is usually the culprit. What happens is the algae (GW form) will flourish off of the ammonia/ammonium and phosphate, remembering that algae can consume phosphate easier than plants because of their thin cell walls, the algae uses up the ammonia/ammonium and phosphate, but it doesn't go away . . . because algae can quickly switch which nutrient it scavenges . . . it moves to nitrates. So you can see why water changes will not rid a tank of GW. Nutrients can be reduced very low in GW and fairly quickly by the GW algaes, but they can scavenge other nutrients . . . iron and trace elements. So, it's very common for the GW to solve the situation that causes it to begin with, but that won't eliminate the GW, for the reasons I've alluded to. Five methods exist to eliminate GW. Blackout, Diatom Filtering, UV Sterilization, Live Daphnia, and Chemical algaecides/flocculents. The first four cause no harm to fish, the fifth one does.

Method No. 1 - Blackout

The blackout method consists of covering the tank for 4 days, so no light whatsoever is allowed into the tank during this time. Cover the tank completely with blankets or black plastic trash bags. Be prepared, killing the algae will result in dead decaying algae that will decompose and pollute the water. Water changes are needed at the beginning and end of the blackout time and ammonia should be monitored also.

Method No. 2 - Diatom Filtering

Diatom filters can sometimes be rented from your LFS. This is a preferred method. Personally, use Magnum 350 w/ Micron Cartridge coated with diatom powder. Diatom filtering removes the algae and doesn't allow it to decay in the tank. You do have to check the filter often, if you have a really bad case of GW the filter can clog pretty quick. Just clean it and start it up again. Crystal clear water usually takes from a few minutes to a couple of hours.

Method No. 3 - UV Sterilization

UV Sterilizers will kill free floating algae. They also kill free floating parasites and bacteria. They also can be problematic for extended use in a planted tank, as they will cause the "breakdown" of some important nutrients. They are expensive and don't remove the decaying material from the tank, if you can afford to keep one they are handy to have around, though not as useful IMO as a diatom filter.

Method No. 4 - Live Daphnia

Adding live daphnia to your tank. This can be a bit tricky. First you need to ensure that you are not adding other "pests" along with the daphnia. Second, unless you can separate the daphnia from the fish, the fish will likely consume the daphnia before the daphnia can consume all the green water.

Method No. 5 - Chemical algaecides/flocculents

The flocculents stick to the gills of fish. Although it won't actually kill them, it does compromise their gill function for quite a while, leaving them open for other maladies

GREEN SPOT ALGAE

Enjoys plenty of light. It forms green spots on aquarium glass and slow growing plants that are exposed to strong light. This algae will appear if CO_2 and Phosphate (PO_4) levels are low. Since it is very hard, algae eaters can't do much in eliminating this algae. *Neritina* snails are the only algae eaters that will graze on Green Spot Algae. It can be scraped manually off the glass with a razor blade or plastic card. In the case of an acrylic aquarium use plastic only. This algae is considered normal in small amounts. To prevent this algae, do weekly water changes, do not overfeed nor over-stock. In planted tanks keep slow growing plants in places where they will get less light (low light set-ups should not suffer from this algae in large extents) and keep Phosphate levels between 0.3-0.5ppm and CO_2 levels 30 ppm.

BLUE-GREEN ALGAE

Even though it's commonly called blue-green *algae* (BGA), it's not classified anymore as one. This "algae" is actually cyanobacteria, a form of life that has both animal and plant characteristics. It forms slimy, blue-green, sheets that will cover everything in a short time and give off a strong, characteristic scent. If left to over-run the tank, cyanobacteria may kill plants and even fish. It doesn't stick and can be easily removed manually, but will return quickly if the underlying water quality issue is not fixed. It can be treated with Erythromycin and other antibiotics, but this method should be done carefully since it might affect the nitrifying bacteria in the gravel and filter, and improper use of antibiotics always brings the risk of developing a more resistant strain. When the BGA gets killed by the algaecide it will start to rot and through that process it will reduce Oxygen levels in the tank. Since the nitrifying bacteria needs O_2 to transfer ammonia/ nitrites into nitrates the nitrifying process will slow down. If algaecide is used, make sure to test the ammonia/nitrite levels. Remove all the visible algae to prevent it from rotting in side the tank.

Some aquarists use the black-out method previously described, where black bags are wrapped around the tank for 4 days and held in complete darkness. It is advisable to raise NO_3 levels to 10-20 ppm before starting the black-out period. Manually remove as much BGA as you can before the blackout, and dead matter after the blackout.Egeria densa (Elodea) and Ceratophyllum demersum are good plants to have in a tank, since these plants are known to secrete natural antibiotic substances that can help prevent BGA. Establishing lots of healthy, fast-growing plants from the day you start the tank, dosing the nitrate levels to maintain 10-20 ppm, and vacuuming the gravel to keep the tank free of decaying matter is the best way to prevent this "algae". BGA can be found in aquariums with very low nitrates because it can fix atmospheric nitrogen. BGA seems not to like high CO_2 levels and stronger water currents.

Green-slimy, Water Surface Film

This is a protein film. Taking a sample under the microscope would probably give some answers. The best way in combating slimy surface film is to improve surface agitation. Filter should be rinsed every week, so the water can flow better through it. To prevent it, use Active Carbon every 2-3 month for 3 weeks and Zeolite to get rid of the organics that tend to build up in aquariums, especially the older ones. 50% weekly water changes will help a lot in reducing organics. The film can be removed with paper towels. Black Mollies will eat this sort of algae. The aquarium shown in the photos are of a low light tank.

CLADOPHORA ALGAE

Cladophora is a branching, green filamentous alga, that forms a moss like structure. This algae doesn't appear to be slimy. Threads are very strong and very thin. It grows on rocks and submersed wood exposed to direct light, and in extreme cases will grow on plants also. Usually it tends to stay on one spot, which makes it easy to remove. Comb it and dose more CO_2. In a case where Cladophora takes over the grassy plants, mow the plants like the lawn. No algae eater is known to eat this kind of algae.

Thread Algae

Grows on leaf edges as individual, up to 30 cm long threads. It is easily removed by twirling a tooth-brush around it. Excess iron is a possible reason. It is better to use substrate iron fertilisers, since this algae uptakes the iron from the water. Healthy plants will out-compete this algae. Algae eaters like SAE and Caridina japonica will consume it. Thread algae is very likely to appear together with the Hair algae.

Forms around the base of slower growing plants, on gravel and bog-wood. It has green-gray color. It grows up to 4 cm, sometimes more. It is easy to remove this algae by twirling a tooth-brush around it. Most aquarists find this algae very welcome as a good food supplement for their fish. Most omnivorous fish like Angels or Barbs will supplement their diet with hair algae if not over-fed. In stronger water currents this algae forms matted clumps, as well as that, stronger water current will disturb their growth. All algae eaters will be more than happy to look after the Hair algae.

Photo credit: Dusko Bojic

Staghorn Algae grows in long individual strands that form a few branches. It will grow close to the light source on equipment and plants. Strands can be pulled off the surface or in very bad cases the whole leaf should be discarded. Higher ammonia/ammonium levels (from overstocking or substrate disturbance) and low CO_2 levels will favour this algae. It's been known that the Siamese Algae Eater will keep this algae in check. Nutrient control and healthy growing plants will limit Staghorn algae.

Beard Algae can actually be an attractive addition to an aquarium on big pieces of stone and/or bogwood. It forms a thick green carpet over the surface closer to the light source. It is very soft and slippery but it is impossible to be removed mechanically. It can also be seen on slow growing plant leaves. It grows approximately 3 cm and the growth is rapid. The best way to control this algae is with the Siamese Algae Eater. Plecos are known to eat this algae, as well as Rosy barbs and Red Tailed Shark, but you should first check whether these fish are compatible with your other fish and with your tank size. Keeping lights for more than 12 hours a day will trigger this algae, as well as unbalanced nutrients. It will show up in planted tanks with low CO_2 and NO_3 levels. This algae can be found in low and high pH waters. Beard Algae is very common in non-planted aquariums.

Grow on leaves and plant stems not necessarily exposed to strong lights. The affected plants are probably suffering deficiency problems and are leaking nutrients back into the water. This algae is considered normal in small quantities. Aquariums with fish such as Siamese Algae Eater, Otos, Amano shrimp, Bristlenose pleco or Mollies will not suffer from this algae. Balanced nutrients will give head start against the algae.

Photo credit: Gianmarco Bertaccini

(Black-brush algae) has been known to thrive in both acidic and alkaline waters. In hard waters it will form lime tissue (from biogenic decalcification) which makes it harder to be eaten by the only algae eater known to eat this type of algae, the Siamese Algae Eater (SAE). Biogenic decalcification can be prevented by adding CO_2. Healthy fast growing plants will out-compete this feathery-black algae that tends to grow on slow growing plant leaves. When buying new plants, before planting, it's good to soak them into a weak household bleach solution for two minutes. 1 part of plain bleach (don't use bleach that has lemon, orange or any kind of scent) to 20 parts of water. Do not forget to rinse the plants well with clean water before adding them into the aquarium. The only perfect way to combat Brush algae is planting lots of healthy fast growing plants, introducing a few SAE, maintaining CO_2 at 30 ppm, nitrates at 15 ppm and phosphates at 0.5 ppm. Leaves that are badly overtaken should be discarded.

Algae are actually zoospores and are commonly found on aquarium glass. They form a dusty looking, green patchy film and in severe cases can cover the whole aquarium glass. It's not known what actually causes this algae. Intense light is favored by GDA. Scraping it off the glass will not help remove this algae since it stays in the water and will float for 30-90 minutes before attaching it self again to the glass. For some reason those zoospores seem to avoid plants, rocks and wood and always go for the glass. Limiting nutrients will not help fighting this algae, but rather cause problems in planted tanks where plants will be exposed to nutrient deficiency and that condition will just favour other algae types. The best known solution for getting rid of GDA has been proposed by Tom Barr. He claims that this algae should be left alone to grow, without wiping the glass for about 10-20 days. After this period GDA will start forming a thick patchy film that will start falling off the glass. When this starts happening it is good to remove this algae out of the tank. This method should keep this algae at bay.

Chapter–7

PHYTOPLANKTON

INTRODUCTION

Phytoplankton are the autotrophic component of the plankton community. The name comes from the Greek words *phyton*, Most phytoplankton are too small to be individually seen with the unaided eye. However, when present in high enough numbers, they may appear as a green discoloration of the water due to the presence of chlorophyll within their cells (although the actual color may vary with the species of phytoplankton present due to varying levels of chlorophyll or the presence of accessory pigments such as phycobiliproteins, xanthophylls, etc.).

Phytoplankton obtain energy through a process called photosynthesis and must therefore live in the well-lit surface layer (termed the euphotic zone) of an ocean, sea, lake, or other body of water. Through photosynthesis, phytoplankton are responsible for much of the oxygen present in the Earth's atmosphere – half of the total amount produced by all plant life. Their cumulative energy fixation in carbon compounds (primary production) is the basis for the vast majority of oceanic and also many freshwater food webs (chemosynthesis is a notable exception). As a side note, one of the more remarkable food chains in the ocean – remarkable because of the small number of links – is that of phytoplankton fed on by krill (a type of shrimp) fed on by baleen whales.

Phytoplankton are also crucially dependent on minerals. These are primarily macronutrients such as nitrate, phosphate or silicic acid, whose availability is governed by the balance between the so-called biological pump and upwelling of deep, nutrient-rich waters. However, across large regions of the World Ocean such as the Southern Ocean, phytoplankton are also limited by the lack of the micronutrient iron. This has led to some scientists advocating iron fertilization as a means to counteract the accumulation of human.

While almost all phytoplankton species are obligate photoautotrophs, there are some that are mixotrophic and other, non-pigmented species that are actually heterotrophic (the latter are often viewed as zooplankton). Of these, the best known are dinoflagellate genera such as *Noctiluca* and *Dinophysis*, that obtain organic carbon by ingesting other organisms or detrital material.

Dinoflagellate

The term 'phytoplankton' encompasses all photoautotrophic microorganisms in aquatic food webs. Phytoplankton serve as the base of the aquatic food web, providing an essential ecological function for all aquatic life. However, unlike terrestrial communities, where most autotrophs are plants, phytoplankton are a diverse group, incorporating protistan eukaryotes and both eubacterial and archaebacterial prokaryotes. There are about 5,000 species of marine phytoplankton.[There is uncertainty in how such diversity has evolved in an environment where competition for only a few resources would suggest limited potential for niche differentiation.

In terms of numbers, the most important groups of phytoplankton include the diatoms, cyanobacteria and dinoflagellates, although many other groups of algae are represented. One group, the coccolithophorids, is responsible (in part) for the release of significant amounts of dimethyl sulfide (DMS) into the atmosphere. DMS is converted to sulfate and these sulfate molecules act as cloud condensation nuclei, increasing general cloud cover. In oligotrophic oceanic regions such as the Sargasso Sea or the South Pacific gyre, phytoplankton is dominated by the small sized cells, called picoplankton, mostly composed of cyanobacteria (*Prochlorococcus*, *Synechococcus*) and picoeucaryotes such as *Micromonas*.

AQUACULTURE

Aquaculture is the farming of freshwater and saltwater organisms including molluscs, crustaceans and aquatic plants. Unlike fishing, aquaculture, also known as aquafarming, implies the cultivation of aquatic populations under controlled conditions. Mariculture refers to aquaculture practiced in marine environments. Particular kinds of aquaculture include algaculture (the production of kelp/seaweed and other algae), fish farming, shrimp farming, oyster farming, and the growing of cultured pearls. Particular methods include aquaponics, which integrates fish farming and plant farming.

Workers harvest catfish from the Delta Pride Catfish farms in MississippiAquaculture has been used in China since circa 2500 BC. When the waters lowered after river floods, some fishes, mainly carp, were held in artificial lakes. Their brood were later fed using nymphs and silkworm feces, while the fish themselves were eaten as a source of protein. By a fortunate genetic mutation, this early domestication of carp led to the development of goldfish in the Tang Dynasty.

The Hawaiian people practiced aquaculture by constructing fish ponds (see Hawaiian aquaculture). A remarkable example from ancient Hawaii is the construction of a fish pond, dating from at least 1,000 years ago, at Alekoko. According to legend, it was constructed by the mythical Menehune. The Japanese practiced cultivation of seaweed by providing bamboo poles and, later, nets and oyster shells to serve as anchoring surfaces for spores. The Romans often bred fish in ponds.

The practice of aquaculture gained prevalence in Europe during the Middle Ages, since fish were scarce and thus expensive. However, improvements in transportation during the 19th century made fish easily available and inexpensive, even in inland areas, causing a decline in the practice. When the first North American fish hatchery was constructed on Dildo Island, Newfoundland Canada in 1889, it was the largest and most advanced in the world.

Americans were rarely involved in aquaculture until the late 20th century, but California residents harvested wild kelp and made legal efforts to manage the supply starting circa 1900, later even producing it as a wartime resource.

Actually, there was keen interest in aquaculture in the United States as early as 1859 when Stephen Ainsworth of West Bloomfield, NY began his experiments with brook trout. By 1864 Seth Green had established a commercial fish hatching operation at Caledonia Springs, near Rochester, NY. By 1866, with the involvement of Dr. W. W. Fletcher of Concord Mass, artificial fish hatching operations were under way in both Canada and the United States.

In contrast to agriculture, the rise of aquaculture is a contemporary phenomenon. According to Prof. Carlos M. Duarte about 430 (97%) of the aquatic species presently in culture have been domesticated since the start of the 20th century, and an estimated 106 aquatic species have been domesticated over the past decade. The domestication of an aquatic species typically involves about a decade of scientific research. Current success in the domestication of aquatic species results from the 20th century rise of knowledge on the basic biology of aquatic species and the lessons learned from past success and failure. The stagnation in the world's fisheries and overexploitation of 20 to 30% of marine fish species have provided additional impetus to domesticate marine species, just as overexploitation of land animals provided the impetus for the early domestication of land species.

In the 1960s, the price of fish began to climb, as wild fish capture rates peaked and the human population continued to rise. Today, commercial aquaculture exists on an unprecedented, huge scale. In the 1980s, open-netcage salmon farming also expanded; this particular type of aquaculture technology remains a minor part of the production of farmed finfish worldwide, but possible negative impacts on wild stocks, which have come into question since the late 1990s, have caused it to become a major cause of controversy

ECONOMIC ROLE

Aquacultre instalations in southern Chile.In 2003, the total world production of fisheries product was 132.2 million tonnes of which aquaculture contributed 41.9 million tonnes or about 31% of the total world production. The growth rate of worldwide aquaculture is very rapid (> 10% per year for most species) while the contribution to the total from wild fisheries has been essentially flat for the last decade.

In the US, approximately 90% of all shrimp consumed is farmed and imported In recent years salmon aquaculture has become a major export in southern Chile, especially in Puerto Montt and Quellón, Chile's fastest-growing city.

Aquaculture is an especially important economic activity in China. Between 1980 and 1997, the Chinese Bureau of Fisheries reports, aquaculture harvests grew at an annual rate of 16.7 percent, jumping from 1.9 million to nearly 23 million tons. In 2005 China accounted for 70% of the world's aquaculture production

ENVIRONMENTAL IMPACTS

Please help improve this section by expanding it. Further information might be found on the talk page or at requests for expansion.

The concentrated nature of aquaculture often leads to higher than normal levels of fish waste in the water. Fish waste is organic and composed of nutrients necessary in all components of aquatic food webs. In some instances such as nearshore, high-intensity operations, increased waste can adversely affect the environment by decreasing dissolved oxygen levels in the water column. Onshore recirculating aquaculture systems, facilities using polyculture techniques, and properly-sited facilities (e.g. offshore or areas with strong currents) are examples of ways to reduce or eliminate the negative environmental effects of fish waste.

Aquaculture can be more environmentally damaging than exploiting wild fisheries Some heavily-farmed species of fish, such as salmon, are maintained in net-contained environments in which the salmon feed exclusively or mostly on wild fish small enough to swim through the net. The salmon consume approximately ten times more energy in fish as they are worth at harvest, making this kind of aquaculture less energy efficient than properly managed fishing.

Despite the environmental concerns, aquaculture profitability is so high that money can and should go back into promoting sustainable practices. urthermore, new methods minimize the risk of biological and chemical pollution through minimizing stress to fish, vaccinating fish, fallowing netpens, and applying Integrated Pest Management. Vaccines also reduce antibiotic use, which are being used more and more.

TYPES OF AQUACULTURE

An open pond Spirulina farmAlgaculture is a form of aquaculture involving the farming of species of algae. The majority of algae that are intentionally cultivated fall into the category of microalgae, also referred to as phytoplankton, microphytes, or planktonic algae.

Macroalgae, commonly know as seaweed, also have many commercial and industrial uses, but due to their size and the specific requirements of the environment in which they need to grow, they do not lend themselves as readily to cultivation on a large scale as microalgae and are most often harvested wild from the ocean.

Fish Farming

Fish farming is the principal form of aquaculture, while other methods may fall under mariculture. It involves raising fish commercially in tanks or enclosures, usually for food. A facility that releases juvenile fish into the wild for recreational fishing or to supplement a species' natural numbers is generally referred to as a fish hatchery. Fish species raised by fish farms include salmon, catfish, tilapia, cod, carp, trout and others. Increasing demands on wild fisheries by commercial fishing operations have caused widespread overfishing. Fish farming offers an alternative solution to the increasing market demand for fish and fish protein.

Freshwater Prawn Farming

A freshwater prawn farm is an aquaculture business designed to raise and produce freshwater prawn or shrimp for human consumption. Freshwater prawn farming shares many characteristics with, and many of the same problems as, marine shrimp farming. Unique problems are introduced by the developmental life cycle of the main species (the giant river prawn, *Macrobrachium rosenbergii*). The global annual production of freshwater prawns (excluding crayfish and crabs) in 2003 was about 280,000 tons, of which China produced some 180,000 tons, followed by India and Thailand with some 35,000 tons each. Additionally, China produced about 370,000 tons of Chinese river crab (*Eriocheir sinensis*)

Integrated Multi-trophic Aquaculture

Integrated Multi-Trophic Aquaculture (IMTA) is a practice in which the by-products (wastes) from one species are recycled to become inputs (fertilizers, food) for another. Fed aquaculture (e.g. fish, shrimp) is combined with inorganic extractive (e.g. seaweed) and organic extractive (e.g. shellfish) aquaculture to create balanced systems for environmental sustainability (biomitigation), economic stability (product diversification and risk reduction) and social acceptability (better management practices).

"Multi-Trophic" refers to the incorporation of species from different trophic or nutritional levels in the same system. This is one potential distinction from the age-old practice of aquatic polyculture, which could simply be the co-culture of different fish species from the same trophic level. In this case, these organisms may all share the same biological and chemical processes, with few synergistic benefits, which could potentially lead to significant shifts in the ecosystem. Some traditional polyculture systems may, in fact, incorporate a greater diversity of species, occupying several niches, as extensive cultures (low intensity, low management) within the same pond. The "Integrated" in IMTA refers to the more intensive cultivation of the different species in proximity of each other, connected by nutrient and energy transfer through water, but not necessarily right at the same location.

Ideally, the biological and chemical processes in an IMTA system should balance. This is achieved through the appropriate selection and proportions of different species providing different ecosystem functions. The co-cultured species should be more than just biofilters; they should also be harvestable crops of commercial value. A working IMTA system should result in greater production for the overall system, based on mutual benefits to the co-cultured species and improved ecosystem health, even if the individual production of some of the species is lower compared to what could be reached in monoculture practices over a short term period.

Sometimes the more general term "Integrated Aquaculture" is used to describe the integration of monocultures through water transfer between organisms For all intents and purposes however, the terms "IMTA" and "integrated aquaculture" differ primarily in their degree of descriptiveness. These terms are sometimes interchanged. Aquaponics, fractionated aquaculture, IAAS (integrated agriculture-aquaculture systems), IPUAS (integrated peri-urban-aquaculture systems), and IFAS (integrated fisheries-aquaculture systems) may also be considered variations of the IMTA concept.

Mariculture

Mariculture is a specialized branch of aquaculture involving the cultivation of marine organisms for food and other products in the open ocean, an enclosed section of the ocean, or in tanks, ponds or raceways which are filled with seawater. An example of the latter is the farming of marine fish, prawns, or oysters in saltwater ponds. Non-food products produced by mariculture include: fish meal, nutrient agar, jewelries (e.g. cultured pearls), and cosmetics.

Shrimp Farming

A shrimp farm is an aquaculture business for the cultivation of marine shrimp for human consumption. Commercial shrimp farming began in the 1970s, and production grew steeply, particularly to match the market demands of the U.S., Japan and Western Europe. The total global production of farmed shrimp reached more than 1.6 million tonnes in 2003, representing a value of nearly 9,000 million U.S. dollars. About 75% of farmed shrimp is produced in Asia, in particular in China and Thailand. The other 25% is produced mainly in Latin America, where Brazil is the largest producer. The largest exporting nation is Thailand.

Shrimp farming has changed from traditional, small-scale businesses in Southeast Asia into a global industry. Technological advances have led to growing shrimp at ever higher densities, and broodstock is shipped worldwide. Virtually all farmed shrimp are penaeids (i.e., shrimp of the family Penaeidae), and just two species of shrimp—the Penaeus vannamei (Pacific white shrimp) and the Penaeus monodon (giant tiger prawn)—account for roughly 80% of all farmed shrimp. These industrial monocultures are very susceptible to diseases, which have caused several regional wipe-outs of farm shrimp populations. Increasing ecological problems, repeated disease outbreaks, and pressure and criticism from both NGOs and consumer countries led to changes in the industry in the late 1990s and generally stronger regulation by governments. In 1999, a program aimed at developing and promoting more sustainable farming practices was initiated, including governmental bodies, industry representatives, and environmental organisations.

Chapter–8

LENTIC ECOSYSTEMS

INTRODUCTION

Lentic ecosystems are the ecosystems of lakes, ponds and swamps. Included in these environments are the biotic interactions (amongst plants, animals and micro-organisms) together with the abiotic interactions (physical and chemical). It refers to standing or still water. It is derived from the Latin *lentus*, which means sluggish. Together with lotic ecosystems, which involves flowing continental waters such as rivers and streams, these fields form the more general study area of freshwater or aquatic ecology.

Lentic systems are diverse, ranging from a small, temporary rainwater pool a few inches deep to Lake Baikal, which has a maximum depth of 1740 m general dist states that ponds and pools have their entire bottom surfaces exposed to light, while lakes do not. In addition, some lakes become seasonally stratified (discussed in more detail below.) Ponds and pools have two regions: the pelagic open water zone, and the benthic zone, which is comprised of the bottom and shore regions. Since lakes have deep bottom regions not exposed to light, these systems have an additional zone, the profundal. These three areas can have very different abiotic conditions and, hence, host species that are specifically adapted to live there.

LENTIC SYSTEM BIOTA

Bacteria

Bacteria are present in all regions of lentic waters. Free-living forms are associated with decomposing organic material, biofilm on the surfaces of rocks and plants, suspended in the water column, and in the sediments of the benthic and profundal zones. Other forms are also associated with the guts of lentic animals as parasites or in commensal relationships Bacteria play an important role in system metabolism through nutrient recycling, which will be discussed in the Trophic Relationships section.

PRIMARY PRODUCERS

Algae, including both phytoplankton and periphyton are the principle photosynthesizers in ponds and lakes. Phytoplankton are found drifting in the water column of the pelagic zone. Many species have a higher density than water which should making them sink and end up in the benthos. To combat this, phytoplankton have developed density changing mechanisms, by forming vacuoles and gas vesicles or by changing their shapes to induce drag, slowing their descent. A very sophisticated adaptation utilized by a small number of species is a tail-like flagella that can adjust vertical position and allow movement in any direction.

Phytoplankton can also maintain their presence in the water column by being circulated in Langmuir rotations Periphytic algae, on the other hand, are attached to a substrate. In lakes and ponds, they can cover all benthic surfaces. Both types of plankton are important as food sources and as oxygen providers. Plants, or macrophytes, in lentic systems live in both the benthic and pelagic zones and can be grouped according to their manner of growth:

1. emergent macrophytes = rooted in the substrate but with leaves and flowers extending into the air;
2. floating-leaved macrophytes = rooted in the substrate but with floating leaves;

3. submersed macrophytes = not rooted in the substrate and floating beneath the surface; and

4. free-floating macrophytes = not rooted in the substrate and floating on the surface.

These various forms of macrophytes generally occur in different areas of the benthic zone, with emergent vegetation nearest the shoreline, then floating-leaved macrophytes, followed by submersed vegetation. Free-floating macrophytes can occur anywhere on the system's surface.

Aquatic plants are more buoyant than their terrestrial counterparts because freshwater has a higher density than air. This makes structural rigidity unimportant in lakes and ponds (except in the aerial stems and leaves). Thus, the leaves and stems of most aquatic plants use less energy to construct and maintain woody tissue, investing that energy into fast growth instead n order to contend with stresses induced by wind and waves, plants must be both flexible and tough. Light is the most important factor controlling the distribution of submerged aquatic plants. Macrophytes are sources of food, oxygen, and habitat structure in the benthic zone, but cannot penetrate the depths of the euphotic zone and hence are not found there.

INVERTEBRATES

Water striders are predatory insects which rely on surface tension to walk on top of water. They live on the surface of ponds, marshes, and other quiet waters. They can move very quickly, up to 1.5 m/s.

Zooplankton are tiny animals suspended in the water column. Like phytoplankton, these species have developed mechanisms that keep them from sinking to deeper waters, including drag-inducing body forms and the active flicking of appendages such as antennae or spines a in the water column may have its advantages in terms of feeding, but this zone's lack of refugia leaves zooplankton vulnerable to predation. In response, some species, especially *Daphnia* sp., make daily vertical migrations in the water column by passively sinking to the darker lower depths during the day and actively moving towards the surface during the night. Also, because conditions in a lentic system can be quite variable across seasons, zooplankton have the ability to switch from laying regular eggs to resting eggs when there is a lack of food, temperatures fall below 2° C, or if predator abundance is high. These resting eggs have a diapause, or dormancy period that should allow the zooplankton to encounter conditions that are more favorable to survival when they finally hatch The invertebrates that inhabit the benthic zone are numerically dominated by small species and are species rich compared to the zooplankton of the open water. They include Crustaceans (e.g. crabs, crayfish, and shrimp), molluscs (e.g. clams and snails), and numerous types of insects. ese organisms are mostly found in the areas of macrophyte growth, where the richest resources, highly oxygenated water, and warmest portion of the ecosystem are found. The structurally diverse macrophyte beds are important sites for the accumulation of organic matter, and provide an ideal area for colonization. The sediments and plants also offer a great deal of protection from predatory fishes.

Very few invertebrates are able to inhabit the cold, dark, and oxygen poor profundal zone. Those that can are often red in color due to the presence of large amounts of hemoglobin, which greatly increases the amount of oxygen carried to cells. Because the concentration of oxygen within this zone is low, most species construct tunnels or borrows in which they can hide and make the minimum movements necessary to circulate water through, drawing oxygen to them without expending much energy.

Fishes and Other Vertebrates

Fishes have a range of physiological tolerances that are dependent upon which species they belong to. They have different lethal temperatures, dissolved oxygen requirements, and spawning needs that are based on their activity levels and behaviors. Because fishes are highly mobile, they are able to deal with unsuitable abiotic factors in one zone by simply moving to another. A detrital feeder in the profundal zone, for example, that finds the oxygen concentration has dropped too low may feed closer to the benthic zone. A fish might also altei its residence during different parts of its life history: hatching in a sediment nest, then moving to the weedy benthic zone to develop in a protected environment with food resources, and finally into the pelagic zone as an adult.

Other vertebrate taxa inhabit lentic systems as well. These include amphibians (e.g. salamanders and frogs), reptiles (e.g. snakes, turtles, and alligators), and a large number of waterfowl species Most of these vertebrates spend part of their time in terrestrial habitats and thus are not directly affected by abiotic factors in the lake or pond. Many fish species are important as consumers and as prey species to the larger vertebrates mentioned above.

TROPHIC RELATIONSHIPS

Lentic systems gain most of their energy from photosynthesis performed by aquatic plants and algae. This autochthonous process involves the combination of carbon dioxide, water, and solar energy to produce carbohydrates and dissolved oxygen. Within a lake or pond, the potential rate of photosynthesis generally decreases with depth due to light attenuation. Photosynthesis, however, is often low at the top few millimeters of the surface, likely due to inhibition by ultraviolet light. The exact depth and photosynthetic rate measurements of this curve are system specific and depend upon:

1. the total biomass of photosynthesizing cells;
2. the amount of light attenuating materials; and
3. the abundance and frequency range of light absorbing pigments (i.e. chlorophylls) inside of photosynthesizing cells. The energy created by these primary producers is important for the community because it is transferred to higher trophic levels via consumption.

Bacteria

The vast majority of bacteria in lakes and ponds obtain their energy by decomposing vegetation and animal matter. In the pelagic zone, dead fisheys and the occasional allochthonous input of litterfall are examples of coarse particulate organic matter (CPOM>1 mm). Bacteria degrade these into fine particulate organic matter (FPOM<1 mm) and then further into usable nutrients. Small organisms such as plankton are also characterized as FPOM. Very low concentrations of nutrients are released during decomposition because the bacteria are utilizing them to build their own biomass. Bacteria, however, are consumed by protozoa, which are in turn consumed by zooplankton, and then further up the trophic levels. Nutrients, including those that contain carbon and phosphorus, are reintroduced into the water column at any number of points along this food chain via excretion or organism death, making them available again for bacteria. This regeneration cycle is known as the microbial loop and is a key component of lentic food webs.

The decomposition of organic materials can continue in the benthic and profundal zones if the matter falls through the water column before being completely digested by the pelagic bacteria. Bacteria are found in the greatest abundance here in sediments, where they are typically 2-1000 times more prevalent than in the water column.

Invertebrates can be divided into feeding guilds based on their method of prey capture. Zooplankton rely on a combination of dissolved organic matter and particulates for survival. Zooplankton concentrate unicellular algae, bacteria, and detritus in the pelagic zone by sieving through morphological structures or with the creation of rotary currents that can centrifuge particles into a larger mass for digestion. These processes are energetically efficient, but they do not allow for the selection of usable material. As a result, a portion of the catch is invaluable as a food source and must be discarded. When capturing larger particles, including other zooplankton taxa, each particle is procured individually with a raptorial appendage. This type of feeding is energetically more costly, but the nutritional returns can be greater because selection is involved.

Benthic invertebrates, due to their high level of species richness, have many methods of prey capture. Filter feeders create currents via siphons or beating cilia, to pull water and its nutritional contents, towards themselves for straining. Grazers use scraping, rasping, and shredding adaptations to feed on periphytic algae and macrophytes. Members of the collector guild browse the sediments, picking out specific particles with raptorial appendages. Deposit feeding invertebrates indiscriminately consume sediment, digesting any organic material it contains. Finally, some invertebrates belong to the predator guild, capturing and consuming living animals. The profundal zone is home to a unique group of filter feeders that use small body movements to draw a current through burrows that they have created in the sediment. This mode of feeding requires the least amount of motion, allowing these species to conserve energy. small number of invertebrate taxa are predators in the profundal zone. These species are likely from other regions and only come to these depths to feed. The vast majority of invertebrates in this zone are deposit feeders, getting their energy from the surrounding sediments.

Fish

Fish size, mobility, and sensory capabilities allow them to exploit a broad prey base, covering multiple zonation regions. Like invertebrates, fish feeding habits can be categorized into guilds. In the pelagic zone, herbivores graze on periphyton and macrophytes or pick phytoplankton out of the water column. Carnivores include fishes that feed on zooplankton in the water column (zooplanktivores), insects at the water's surface, on benthic structures, or in the sediment (insectivores), and those that feed on other fishes (piscivores). Fish that consume detritus and gain energy by processing its organic material are called detritivores. Omnivores ingest a wide variety of prey, encompassing floral, faunal, and detrital material. Finally, members of the parasitic guild acquire nutrition from a host species, usually another fish or large vertebrate Fish taxa are flexible in their feeding roles, varying their diets with environmental conditions and prey availability. Many species also undergo a diet shift as they develop. Therefore, it is likely that any single fish occupies multiple feeding guilds within its lifetime.

Lentic Food Webs

As noted in the previous sections, the lentic biota are linked in complex web of trophic relationships. These organisms can be considered to loosely be associated with specific trophic groups (e.g. primary producers, herbivores, primary carnivores, secondary carnivores, etc.). Scientists have developed several theories in order to understand the mechanisms that control the abundance and diversity within these groups. Very generally, top-down processes dictate that the abundance of prey taxa is dependent upon the actions of consumers from higher trophic levels. Typically, these processes operate only between two trophic levels, with no effect on the others. In some cases, however, aquatic systems experience a trophic cascade; for example, this might occur if primary producers experience less grazing by herbivores because these herbivores are suppressed by carnivores. Bottom-up processes are functioning when the abundance or diversity of members of higher trophic levels is dependent upon the availability or quality of resources from lower levels. Finally, a combined regulating theory, bottom-up:top-down, combines the predicted influences of consumers and resource availability. It predicts that trophic levels close to the lowest trophic levels will be most influenced by bottom-up forces, while top-down effects should be strongest at top levels.

LOCAL SPECIES RICHNESS

The biodiversity of a lentic system increases with the surface area of the lake or pond. This is attributable to the higher likelihood of partly terrestrial species of finding a larger system. Also, because larger systems typically have larger populations, the chance of extinction is decreased. Additional factors, including temperature regime, pH, nutrient availability, habitat complexity, speciation rates, competition, and predation, have been linked to the number of species present within systems.

Phytoplankton and zooplankton communities in lake systems undergo seasonal succession in relation to nutrient availability, predation, and competition. Sommer *et al.*

described these patterns as part of the Plankton Ecology Group (PEG) model, with 24 statements constructed from the analysis of numerous systems. The following includes a subset of these statements, as explained by Brönmark and Hansson illustrating succession through a single seasonal cycle:

Winter

1. Increased nutrient and light availability result in rapid phytoplankton growth towards the end of winter. The dominant species, such as diatoms, are small and have quick growth capabilities.
2. These plankton are consumed by zooplankton, which become the dominant plankton taxa.

Spring

3. A clear water phase occurs, as phytoplankton populations become depleted due to increased predation by growing numbers of zooplankton.

Summer

4. Zooplankton abundance declines as a result of decreased phytoplankton prey and increased predation by juvenile fishes.
5. With increased nutrient availability and decreased predation from zooplankton, a diverse phytoplankton community develops.
6. As the summer continues, nutrients become depleted in a predictable order: phosphorus, silica, and then nitrogen. The abundance of various phytoplankton species varies in relation to their biological need for these nutrients.
7. Small-sized zooplankton become the dominant type of zooplankton because they are less vulnerable to fish predation.

Fall

8. Predation by fishes is reduced due to lower temperatures and zooplankton of all sizes increase in number.

Winter

9. Cold temperatures and decreased light availability result in lower rates of primary production and decreased phytoplankton populations.
10. Reproduction in zooplankton decreases due to lower temperatures and less prey.

The PEG model presents an idealized version of this succession pattern, while natural systems are known for their variation

Latitudinal Patterns

There is a well-documented global pattern that correlates decreasing plant and animal diversity with increasing latitude, that is to say, there are fewer species as one moves towards the poles. The cause of this pattern is one of the greatest puzzles for ecologists today. Theories for its explanation include energy availability, climatic variability, disturbance, competition, etc. Despite this global diversity gradient, this pattern can be weak for freshwater systems compared to global marine and terrestrial systems. found that smaller organisms (protozoa and plankton) did not follow the expected trend strongly, while larger species (vertebrates) did. They attributed this to better dispersal ability by smaller organisms, which may result in high distributions globally.

Natural Lake Lifecycles

Lakes can be formed in a variety of ways, but the most common are discussed briefly below. The oldest and largest systems are the result of tectonic activities. The rift lakes in Africa, for example are the result of seismic activity along the site of separation of two tectonic plates. Ice-formed lakes are created when glaciers recede, leaving behind abnormalities in the landscape shape that are then filled with water. Finally, oxbow lakes are fluvial in origin, resulting when a meandering river bend is pinched off from the main channel.

All lakes and ponds receive sediment inputs. Since these systems are not really expanding, it is logical to assume that they will become increasingly shallower in depth, eventually becoming wetlands or terrestrial vegetation. The length of this process should depend upon a combination of depth and sedimentation rate. Mos ives the example of Lake Tanganyika, which reaches a depth of 1500 m and has a sedimentation rate of 0.5 mm/yr. Assuming that sedimentation is not influenced by anthropogenic factors, this system should go extinct in approximately 3 million years. Shallow lentic systems might also fill in as swamps encroach inward from the edges. These processes operate on a much shorter timescale, taking hundreds to thousands of years to complete the extinction process

Acidification

Sulfur dioxide and nitrogen oxides are naturally released from volcanoes, organic compounds in the soil, wetlands, and marine systems, but the majority of these compounds come from the combustion of coal, oil, gasoline, and the smelting of ores containing sulfur. These substances dissolve in atmospheric moisture and enter lentic systems as acid rain Lakes and ponds that contain bedrock that is rich in carbonates have a natural buffer, resulting in no alteration of pH. Systems without this bedrock, however, are very sensitive to acid inputs because they have a low neutralizing capacity, resulting in pH declines even with only small inputs of acid. At a pH of 5-6 algal species diversity and biomass decrease considerably, leading to an increase in water transparency – a characteristic feature of acidified lakes. As the pH continues lower, all fauna becomes less diverse. The most significant feature is the disruption of fish reproduction. Thus, the population is eventually composed of few, old individuals that eventually die and leave the systems without fishes. Acid rain has been especially harmful to lakes in the Northeastern United States.

Eutrophication

Eutrophic systems contain a high concentration of phosphorus (~30+µg/L), nitrogen (~1500+µg/L), or bothPhosphorus enters lentic waters from wastewater treatment effluents, discharge from raw sewage, or from runoff of farmland. Nitrogen mostly comes from agricultural fertilizers from runoff or leaching and subsequent groundwater flow. This increase in nutrients required for primary producers results in a massive increase of phytoplankton growth, termed a plankton bloom. This bloom decreases water transparency, leading to the loss of submerged plants. The resultant reduction in habitat structure has negative impacts on the species' that utilize it for spawning, maturation and general survival. Additionally, the large number of short-lived phytoplankton result in a massive amount of dead biomass settling into the sediment. Bacteria need large amounts of oxygen to decompose this material, reducing the oxygen concentration of the water. This is especially pronounced in stratified lakes when the thermocline prevents oxygen rich water from the surface to mix with lower levels. Low or anoxic conditions preclude the existence of many taxa that are not physiologically tolerant of these conditions

Invasive species have been introduced to lentic systems through both purposeful events (e.g. stocking game and food species) as well as unintentional events (e.g. in ballast water). These organisms can affect natives via competition for prey or habitat, predation, habitat alteration, hybridization, or the introduction of harmful diseases and parasites.With regard to native species, invaders may cause changes in size and age structure, distribution, density, population growth, and may even drive populations to extinction Examples of prominent invaders of lentic systems include the zebra mussel and sea lamprey in the Great Lakes.

DEEP LAKE WATER COOLING

Deep lake water cooling uses cold water pumped from the bottom of a lake as a heat sink for climate control systems. Because heat pump efficiency improves as the heat sink gets colder, deep lake water cooling can reduce the electrical demands of large cooling systems where it is available. It is similar in concept to modern geothermal sinks, but generally simpler to construct given a suitable water source.

Water is most dense at 3.98 °C at standard atmospheric pressure. Thus as water cools below 3.98 °C it lowers in density and will rise, the most obvious example being that ice floats. As the temperature climbs above 3.98 °C, water density also decreases and causes the water to rise, which is why lakes are warmer on the surface during the summer. The combination of these two effects means that the bottom of most deep bodies of water located well away from the equatorial regions is at a constant 3.98° C.

Air conditioners are heat pumps. During the summer, when outside air temperatures are higher than the temperature inside a building, air conditioners use electricity to transfer heat from the cooler interior of the building to the warmer exterior ambient. This process uses electrical energy.

Unlike residential air conditioners, most modern commercial air conditioning systems do not transfer heat directly into the exterior air. The thermodynamic efficiency of the overall system can be improved by utilizing evaporative cooling, where the temperature of the cooling water is lowered close to the wet-bulb temperature by evaporation in a cooling tower. This cooled water then acts as the heat sink for the heat pump.

Deep lake water cooling allows an even higher thermodynamic efficiency by utilizing the deep lake water, which is at a lower heat rejection temperature than the ambient wet bulb temperature. The higher efficiency results in less electricity used. For many buildings, the lake water is sufficiently cold that the refrigeration portion of the air conditioning systems can be shut down during some environmental conditions and the building interior heat can be transferred directly to the lake water heat sink. This is referred to as "free cooling", but is not actually free, since pumps and fans run to circulate the lake water and building air.

One added attraction of deep lake water cooling is that it saves energy during peak load times, such as summer afternoons, when a sizable amount of the total electrical grid load is air conditioning.

Cornell University's Lake Source Cooling System uses Cayuga Lake as a heat sink to operate the central chilled water system for its campus and to also provide cooling to the Ithaca City School District. The system has operated since the summer of 2000 and was built at a cost of $55-60 million. It cools a 14,500 ton (51 megawatt) load.

Lake water enters the system via a screened intake structure 10,400 feet (3,200 m) away in 250 feet (76 m) of water. The intake pipeline is 63-inch (1.6 m) High Density Polyethylene (HDPE) that was deployed from the surface using a "controlled" sink process where water was pumped in at the shallow end and air was released at the other end. A series of stiffener rings and concrete collars keep the pipeline on the lake floor and protect it from mechanical forces. The outfall is 48-inch (1,200 mm) HDPE and is approximately 750 feet (230 m) long. The last 100 feet (30 m) of the outfall has 38 six-inch (152 mm) nozzles, about 1-foot (0.30 m) above the bottom of the lake floor in 14 feet (4.3 m) of water, pointed up at a 20 degree angle and pointed north only. This helps promote mixing of the return water into the receiving water. The water cools a heat-exchanger which is connected to a closed-loop campus chilled water distribution system linked to many buildings on the main campus.

Since August 2004, a deep lake water cooling system has been operated by the Enwave Energy Corporation in Toronto, Ontario. It draws water from Lake Ontario through tubes extending 5 km into the lake, reaching to a depth of 83 metres. The deep lake water cooling system is part of an integrated district cooling system that covers Toronto's financial district, and has a cooling power of 59,000 tons (207 MW). The system currently has enough capacity to cool 3.2 million square meters of office space

The cold water drawn from Lake Ontario's deep layer in the Enwave system is not returned directly to the lake, once it has been run through the heat exchange system. The Enwave system only uses water that is destined to meet the city's domestic water needs. Therefore, the Enwave system does not pollute the lake with a plume of waste heat.

Ocean Water Cooling

The Inter-continental Resort and Thalasso-Spa on the island of Bora Bora uses a cold seawater system to air-condition its buildings The system accomplishes this by passing cold seawater through a heat exchanger where it cools freshwater in a closed loop system. This cool freshwater is then pumped to buildings and is used for cooling directly (no conversion to electricity takes place).

Comparison to Related Technologies

This water-cooling technology has some relationship to an older technology and a possible future technology. Looking back to the past, water-cooling recalls well insulated icehouses which were used to store ice throughout the year prior to the invention of refrigeration. Icehouses were filled with fresh ice collected from lake surfaces during the winter whereas deep lake water cooling taps a permanent store of cold water.

Looking towards the future, water-cooling uses cold deep water just as ocean thermal energy conversion (OTEC) does. However, OTEC is intended to be used for generating energy by operating a heat engine on the temperature difference between the ocean bottom and the ocean surface. Deep lake water cooling bypasses the need for electricity generation altogether and, so, is a simpler and more immediately practical technology than OTEC. Ambitious OTEC projects have yet to realize their full potential because they present far more demanding engineering challenges.

Chapter–9

Alginate or Alginic Acid

INTRODUCTION

Alginic acid (algin, alginate) is a viscous gum that is abundant in the cell walls of brown algae.

Chemically, it is a linear copolymer with homopolymeric blocks of (1-4)-linked ß-D-mannuronate (M) and its C-5 epimer a-L-guluronate (G) residues, respectively, covalently linked together in different sequences or blocks.

The monomers can appear in homopolymeric blocks of consecutive G-residues (G-blocks), consecutive M-residues (M-blocks), alternating M and G-residues (MG-blocks) or randomly organized blocks.

Commercial varieties of alginate are extracted from seaweed, including the giant kelp *Macrocystis pyrifera*, *Ascophyllum nodosum* and various types of *Laminaria*. It is also produced by two bacterial genera *Pseudomonas* and *Azotobacter*, which played a major role in the unravelling of its biosynthesis pathway. Bacterial alginates are useful for the production of micro- or nanostructures suitable for medical applications.

Alginate absorbs water quickly, which makes it useful as an additive in dehydrated products such as slimming aids, and in the manufacture of paper and textiles. It is also used for waterproofing and fireproofing fabrics, as a gelling agent, for thickening drinks, ice cream and cosmetics, and as a detoxifier that can absorb poisonous metals from the blood. Alginate is also produced by certain bacteria, notably Azotobacter species. Alginate ranges from white to yellowish brown, and takes filamentous, grainy, granular, and powdered forms. It is insoluble in water and organic solvents, and dissolves slowly in basic solutions of sodium carbonate, sodium hydroxide and trisodium phosphate.

Alginate is used in various pharmaceutical preparations such as Gaviscon, Bisodol, Asilone. Alginate is used extensively as a mold-making material in dentistry, prosthetics, lifecasting, and in textiles. It is also used in the food industry, for thickening soups and jellies. Calcium alginate is used in different types of medical products, including burn dressings that promote healing and which can be removed with less pain than conventional dressings. Also, due to alginate's biocompatibility and simple gelation with divalent cations such as Ca 2+, it is widely used for cell immobilization and encapsulation.

Alginic acid (alginato) is also used in culinary arts, most notably in the "Esferificación" (Sphereification) techniques of Ferrán Adriá of Barcelona's El Bulli, in which natural juices of fruits and vegetables are encapsulated in bubbles that "explode" on the tongue when consumed. One of the most famous examples of this use of alginic acid is where Ferran Adria used alginic acid to make apple caviar. Due to its ability to absorb water quickly, Alginate can be changed through a lyophilization process to a new structure which has the ability to expand. It is used in the weight loss industry as an appetite suppressant. A new type of diet using Alginate is the CM3-Alginate Diet.

SODIUM ALGINATE

The chemical compound sodium alginate is the sodium salt of alginic acid. Its empirical chemical formula is $NaC_6H_7O_6$. Its form as a gum, when extracted from the cell walls of brown algae, is used by the foods industry to increase viscosity and as an emulsifier. It is also used in indigestion tablets and the preparation of dental impressions. Sodium alginate has no discernible flavour.

Another major use of sodium alginate is reactive dye printing, where it is used in the textile industry. A major application for sodium alginate is as thickener for reactive dyestuffs (such as the Procion cotton-reactive dyes) in textile screen-printing and carpet jet-printing. Alginates do not react with these dyes and wash out easily, unlike starch-based thickeners. Sodium alginate is a good chelator for pulling radioactive toxins such as iodine-131 and strontium-90 from the body which have taken the place of their non-radioactive counterparts It is also used in immobilizing enzymes by inclusion.

As a food additive, sodium alginate is used especially in the production of gel-like foods. For example, bakers' "Chellies" are often gelled alginate "jam." Also, the pimento stuffing in prepared cocktail olives is usually injected as a slurry at the same time that the stone is ejected; the slurry is subsequently set by immersing the olive in a solution of a calcium salt which causes rapid gelation by electrostatic cross-linking. A similar process can be used to make "chunks" of everything from cat food through "reformed" ham or fish to "fruit" pieces for pies. It has the E-number 401. It is also know as a harmless, inexpensive, but effective way to lower cholesterol, and also remove harmfull toxins.

ASCOPHYLLUM NODOSUM

Ascophyllum nodosum is a large, common, brown alga, in the Class Phaeophyceae. It is seaweed of the northern Atlantic Ocean, also known as *Norwegian Kelp*, *Knotted Kelp*, *knotted wrack* or *egg wrack*. It is common on the north-western coast of Europe (from Svalbard to Portugal) including east Greenland and the north-eastern coast of North America.

Ascophyllum is very popular amongst the science community and has been claimed to be both the best known seaweed on the planet as well as the most researched by the academic community.

Ascophyllum nodosum has long fronds with large egg-shaped air-bladders set in series at regular intervals in the fronds and not stalked. The fronds can reach 2 m in length and are attached by a holdfast to rocks and boulders. The fronds are olive-brown in color and somewhat compressed but without a mid-rib.

This seaweed grows quite slowly and can live for several decades; it may take approximately five years before becoming fertile.

Life history is of one diploid plant and gametes. The gametes are produced in conceptacles embedded in yellowish receptacles on short branches.

Ascophyllum nodosum is found mostly on sheltered sites on shores in the mid-littoral where it can become the dominant species in the littoral zone.

Polysiphonia lanosa (L.) Tandy is a small red alga, commonly found growing in dense tufts on *Ascophyllum* whose rhizoids penetrate the host. It is considered by some as parasitic.

Varieties and Forms

Several different varieties and forms of this species have been described.

Ascophyllum nodosum var *minor* has been described from Larne Lough in Northern Ireland.

There are free floating ecads of this species such as *Ascophyllum nodosum mackaii* Cotton, which is found at very sheltered locations, such as at the heads of sea lochs in Scotland and Ireland.

The species is found in a range of coastal habitats from sheltered estuaries to moderately exposed coasts, often it dominates the inter-tidal zone (although sub-tidal populations are known to exist in very clear waters). However it is rarely found on exposed shores, and if it is found the fronds are usually small and badly scratched.

It has been recorded as an accidental introduction to San Francisco, California, and eradicated as a potential invasive species there.

Recorded in Europe from: Faroe Islands, Norway, Ireland, Britain and Isle of Man Netherlands North America: Bay of Fundy, Nova Scotia, Baffin Island, Hudson Strait, Labrador and Newfoundland.

USES

Ascophyllum nodosum is harvested for use in alginates, fertilisers and for the manufacture of seaweed meal for animal and human consumption. It has long been used as an organic and mainstream fertilizer for many varieties of crops due to its combination of both macronutrient, (eg. N, P, K, Ca, Mg, S) and micronutrients (eg. Mn, Cu, Fe, Zn, etc.) It also host to cytokinins, auxin-like, gibberellins, betaines, mannitol, organic acids, polysaccharides, amino acids, and proteins which are all very beneficial and widely used in agriculture.

Ascophyllum nodosum along with *Macrocystis pyfera*is harvested in Ireland, Scotland and Norway from which alginates are extracted it is one of the world's principal alginate supply.

Chapter–10

FILAMENTOUS ALGAE

INTRODUCTION

Fertilization to produce a phytoplankton or algal "bloom" can prevent the establishment of filamentous algae if started early enough in the spring. Fertilization also produces a strong food chain to the pond fish.

Non-toxic dyes or colorants prevent or reduce aquatic plant growth by limiting sunlight penetration, similar to fertilization. Aquashade is an example of non-toxic dye and other products are available. However, dyes do not enhance the natural food chain and may suppress the natural food chain of the pond.

Grass carp will seldom control aquatic vegetation the first year they are stocked. They will consume filamentous algae but is not a preferred food item. Therefore, they will usually consume other types of submerged vegetation before they consume filamentous algae. Grass carp stocking rates that will control filamentous algae are usually between 10 and 20 per surface acre. In Texas, only triploid grass carp are legal and a permit from the Texas Parks and Wildlife Department is required before they can be purchased from a certified dealer.

Tilapia will consume filamentous algae but are a warm water species that cannot survive in temperatures below 55° F. Therefore, tilapia usually cannot be stocked before mid-

April or May and will die in November or December. Recommended stocking rates are 15 to 20 pounds of mixed sex adult Mozambique tilapia (Oreochromis mossambicus) per surface area. Tilapias are often not effective for vegetation control if the pond has a robust bass population due to intense predation. In Texas, stocking of Mozambique tilapia does not require a permit from the Texas Parks and Wildlife Department. Any other species of tilapia would require a permit. Check with out County Extension Agent in other states for legality of stocking tilapia.

Chemical

The active ingredients that have been successful in treating filamentous algae include copper based compounds (E), alkylamine salts of endothall (G), sodium carbonate peroxyhydrate (G), and diquat (G). E = excellent, G = good.

Copper Sulfate or "blue stone" is probably the most commonly used algal treatments because of its availability and low cost. Copper sulfate comes in several forms depending on how finely it is ground. Smaller crystals will dissolve easier than larger crystals. In very hard water it is difficult to use copper sulfate because it binds with the calcium, precipitates out of solution, and renders the copper ineffective as an algaecide.

All copper compounds can be toxic to fish if used above labeled rates and can be toxic in soft or acidic waters even at label rates. Before using copper is it best to test the pond water alkalinity and adjust copper treatments to alkalinity concentrations. For additional information on using copper sulfate see the SRAC #410 Calculating Treatments for Ponds and Tanks.

Cutrine Plus, K-Tea, Captain, and Clearigate are all chelated or compound copper herbicides that are effective on filamentous algae. Other chelated or compound copper formulations are available but are not linked to this web site.

Hydrothol is an alkylamine salt of endothall which can be used to control filamentous algae and comes in liquid and granular formulations. It is a contact algaecide and herbicide. Contact herbicides act quickly and kill all plants cells that they contact. Hydrothol can be toxic to fish.

GreenClean, PAK27, and Phycomycin are Sodium Carbonate Peroxyhydrate based herbicides. These are pelleted contact herbicides for control of blue-green algae. Hydrogen peroxide is the active agent in this algaecide. It is not effective on the macroalgaes, *Chara* or *Nitella*, or on any higher plants.

Reward is the registered diquat label for aquatic use and has been found effective on filamentous algae. It is a contact algaecide and herbicide. Contact herbicides act quickly and kill all plants cells that they contact.

One danger with any chemical control method is the chance of an oxygen depletion after the treatment caused by the decomposition of the dead plant material. Oxygen depletions can kill fish in the pond. If the pond is heavily infested with weeds it may be possible (depending on the herbicide chosen) to treat the pond in sections and let each section decompose for about two weeks before treating another section. Aeration, particularly at night, for several days after treatment may help control the oxygen depletion.

One common problem in using aquatic herbicides is determining area and/or volume of the pond or area to be treated. To assist you with these determinations see SRAC #103 Calculating Area and Volume of Ponds and Tanks.

Many aquatically registered herbicides have water use restrictions. For General Water Use Restrictions click here. To see the labels for these products click on the name. Always read and follow all label directions. Check label for specific water use restrictions. Filamentous algae are seldom encouraged. It could be encouraged by transplanting mats of filamentous algae from another pond and not fertilizing to encourage a planktonic algae bloom.

The most common aquatic weed problem in Ohio ponds is filamentous algae. Its presence can degrade water quality and recreational enjoyment. Excessive algae can cause an oxygen depletion leading to a fish kill when it decomposes as a result of natural die-off or herbicide application. Early and regular control measures will help reduce the problems associated with filamentous algae.

THE PLANT

Filamentous algae, also called "moss" or "pond scum," forms dense mats of hairlike strands. Its growth begins on submerged objects on the pond bottom. As it grows, the algae gives off oxygen that becomes entrapped in the mat of strands. This gives it buoyancy and causes it to rise to the surface where it frequently covers large areas of the pond. Filamentous algae is often a persistent problem because it reproduces by plant fragments, spores and cell division.

There are many species of filamentous algae and microscopic examination is usually required to make an exact identification. However, some of the more common forms can be distinguished by their texture. Spirogyra is bright green and slimy to the touch; Cladophora has a cottony feel; and Pithophora is often referred to as "horse hair" algae because its coarse texture resembles that of horse hair and it may feel like steel wool.

Mechanical Control

Filamentous algae can be controlled by physically removing large floating clumps with a rake. This will prevent the algae from decomposing in the pond and consuming dissolved oxygen. Algae that has been removed can be piled for composting or used in a garden as mulch.

Steepening the sides of the pond to achieve a 3:1 slope will eliminate shallow water areas so that sunlight cannot reach bottom-growing algae. However, if the entire pond has filled in as a result of sedimentation or decaying vegetation, a dragline or dredge may be needed to deepen the pond.

BIOLOGICAL CONTROL

Biological control involves disrupting plant growth by modifying the aquatic environment through natural manipulation, or it can mean introducing a living organism that is capable of controlling aquatic vegetation.

One method of biological control is maintaining a fertility level that fosters the development of a microscopic plant and animal population, which prevents sunlight penetration. This requires intense management and more time than the average pond owner may wish to devote to the pond. Sunlight penetration to the pond bottom where the algae begins to grow can also be reduced by introducing an inert dye (usually blue).

The addition of triploid white amur (a vegetation-eating fish) as a biological control measure may have mixed results. Filamentous algae is not a preferred food, but will be eaten if no other vegetation is present. If other aquatic plants such as water milfoil or coontail are readily available, the filamentous algae may be ignored and continue to flourish.

CHEMICAL CONTROL

Copper Sulfate

Most species of algae can be controlled with very low concentrations of copper sulfate. It is available in crystalline nuggets the size of rock salt or as a finely ground "snow" grade (Figure 1). The recommended treatment rate is 2.7 pounds per acre-foot of water. [Acre-feet is a volume measurement of the pond. It is determined by multiplying average depth (feet) X surface area (acres). For more information on calculating measurements, consult Pond Measurements (Natural Resources Facts A-2), available from county offices of Ohio State University Extension.] When uniformly applied, this will result in a 1 part per million (ppm) concentration throughout the volume of the pond. For very hard water (more than 12 grains or 200 parts per million of hardness), this rate should be doubled.

The method of application will determine what size of copper sulfate crystals to purchase. The important principle to keep in mind is that actual contact of the copper sulfate with the algae is necessary in order to achieve satisfactory control. For best results, dissolve copper sulfate in water and spray it directly on floating algal mats or on the water surface above submerged algae. Finely ground, "snow grade" copper sulfate is best for this method as it dissolves easier. Mix the desired amount of copper sulfate with enough water to cover the area to be treated, and apply with a sprayer or bucket and dipper. Because copper is corrosive to galvanized metal, application equipment and mixing containers should be made of plastic or stainless steel.

In large ponds and when spray equipment is not available, it may be easier to treat with copper sulfate by placing the larger crystals of this chemical in a burlap bag and towing the bag through the water until all the crystals have been dissolved in the area to be treated. One application of copper sulfate is unlikely to provide season-long control. Re-treatment may be necessary at 3-4 week intervals.

There are no water-use restrictions associated with the use of copper sulfate. When applied at the proper rate, the water may be used immediately for swimming, drinking, fishing, irrigation and livestock. However, since copper sulfate has a metallic odor, pond owners may want to suspend drinking, swimming and livestock watering uses for 12 hours.

Copper Chelate

Copper is also available in a chelated, or buffered, formulation, which is manufactured as a liquid or granule. This provides some advantages during application. The liquid form needs only to be mixed with water and sprayed out over the pond surface; there are no crystals to dissolve. The granular formulation consists of a clay granule impregnated with copper chelate. As the granule breaks down, the copper is released into the water. This formulation is especially useful when spot treatment is desirable. Granules are best suited for application early in the growing season because of the time required (2-3 weeks) for them to dissolve and release the chemical. There are no water-use restrictions associated with either formulation of copper chelate.

Diquat Dibromide

This is a contact herbicide that will control some, but not all, species of filamentous algae. It is applied by pouring directly from the container or by diluting with water and injecting below the water surface. For best results, it should be applied before algae growth reaches the surface. Diquat dibromide should not be used in muddy water. There are water-use restrictions associated with this material. Read the label.

Endothall

The amine salt formulation of endothall (sold as Hydrothol 191) is labeled for algae control. It is available as a liquid or granular material. Endothall is a contact herbicide and is most effective in waters 65 degrees F and above. Fish are extremely sensitive to this material. Read the label for water-use restrictions.

Copper-Resistant Algae

One form of filamentous algae, Pithophora, can be especially troublesome because it is resistant to normal applications of copper compounds. Although it is not widespread, scattered reports of Pithophora in Ohio ponds are received every year. If, after a normal treatment with copper sulfate, there is algae remaining that does not appear to be affected, it may be Pithophora. Positive identification can be made by sending a sample to the Plant and Pest Diagnostic Clinic at Ohio State University. Samples can be submitted directly to the clinic or through the county offices of Ohio State University Extension.

Special Precautions

Fish are extremely sensitive to Hydrothol 191. To reduce the hazard of a fish kill, start application at the shoreline and move outward so that fish can escape from treated areas. Select another product if fish toxicity is a concern.

Copper sulfate is corrosive to galvanized containers. Therefore, the solution should be mixed in wooden, earthenware, plastic, stainless steel or copper-lined containers. If a sprayer is not available, you may broadcast the solution with a plastic watering can or bucket and dipper.

If the algae is so abundant that it covers more than half of the total pond surface, a complete treatment may result in an oxygen depletion and fish kill. This hazard is greatest during very hot, overcast weather. When these conditions exist, treat only half the pond and wait 10-14 days before treating the other half.

Copper compounds applied at the recommended rates are lethal to fish eggs and some species of newly hatched fish. These materials should not be applied during spawning periods, unless it is desirable to destroy the eggs and the new hatch. Bass will begin to construct shallow depressions in the pond bottom when the water reaches 60 degrees F. Eggs are deposited by the female and guarded by the male for 3-14 days. Within a couple of weeks after the bass have spawned and when the water temperature reaches 70 degrees F, bluegill and redear sunfish will be seen building nests in the shallow areas. As with the bass, the male guards the nest after the eggs have been deposited. These eggs will hatch in a few days. Bass will only spawn once in the spring, but forage fish (bluegill, redear sunfish and minnows) will spawn throughout much of the summer and some individuals may spawn several times in a single season. To avoid the application of copper compounds during the spawning season, monitor the water temperature and look for active nests in the shallow areas of the pond.

Chapter–11

ALGAE AND MICROORGANISMS

INTRODUCTION

Algae , like *Lyngbya* and other microorganismsExtremely small organism that can only be seen using a microscope., have been with us for thousands of years, and are found around the world in slower moving or still waters of a given temperature, where sunshine and a cocktail of nutrientsAny food, chemical element or compound an organism requires to live, grow, or reproduce. are present for them to grow by photosynthesisIs the chemical process where plants and some bacteria can capture and organically fix the energy of the sun. This chemical reaction can be described by the following simple equation:

$$6CO_2 + 6H_2O + \text{light energy} >>> C_6H_{12}O_6 + 6O_2 \qquad (11.1)$$

The main product of photosynthesis is a carbohydrate, such as the sugar glucose, and oxygen which is released to the atmosphere. All of the sugar produced in the photosynthetic cells of plants and other organisms is derived from the initial chemical combining of carbon dioxide and water with sunlight. This chemical reaction is catalyzed by chlorophyll acting in concert with other pigment, lipid, sugars, protein, and nucleic acid molecules. Sugars created in photosynthesis can be later converted by the plant to starch for storage, or it can be combined with other sugar molecules to form specialized carbohydrates such as cellulose, or it can be combined with other nutrients such as nitrogen, phosphorus, and sulfur, to build complex molecules such as proteins and nucleic acids. Also see chemosynthesis. It is said that photosynthesis gives rise to three quarters of the world supply of oxygen that we breathe. (using energy from the sunshine). There are two basic variants of algae; those having cells with a nucleus and those of a bacterial nature such as Cyanobacteria Bacteria that have the ability to photosynthesize. (blue-green algae) of which *Lyngbya* is an example. The sixty subspecies of *Lyngbya* have both salt water and fresh waterWater that is relatively free of salts attributes.

All the conditions for *Lyngbya* to thrive existed in Crystal River/Kings Bay long ago, just as they do today. The water temperature issuing from the springs was and remains ideal. Florida sunshine was the same then as today, and the requisite cocktail of nutrients was and remains present in the aquiferPorous, water-bearing layers of sand, gravel, or rock. today. In days of old, leachate from pine forests, guano, limestone Sedimentary rock composed of carbonate minerals, especially calcium carbonate. Limestone can be created by clastic and non-clastic processes. Clastic limestones are formed from the break up and deposition of shells, coral and other marine organisms by wave-action and ocean currents. Non-clastic limestones can be formed either as a precipitate or by the lithification of coral reefs, marine organism shells, or marine organism skeletons. rock formations, salt and sandy soil was sufficient source for the nitrogen, phosphorous, dissolved calcium, iron, and heavy metal mineral. A naturally occurring inorganic solid with a crystalline structure and a specific chemical composition. Over 2,000 types of minerals have been classified. traces required for the nutrient cocktail. In some parts of the world they still are – even where man had never influenced the environment unduly.

For example, slash pine plantations for lumber, phosphate mining, limestone mining for cement, citrus and cattle farming and other agriculture all have produced excess nutrient run off into the aquifer and soil, and the more so because natural trees, vegetation and two thirds of our wetlands. Natural land-use type that is covered by salt water or fresh water for some time period. This land type can be identified by the presence of particular plant species or characteristic conditions. were destroyed to make way.

MODERN URBANIZATION

Expansion of cities into rural regions because of population growth. In most cases, population growth is primarily due to the movement of rural based people to urban areas. This is especially true in Less Developed Countries and motorized traffic densities spill many pollutants. Something which contaminates (water, air, etc.) with harmful or poisonous substances counted as parameters of poor water quality. A term used to describe the chemical, physical, and biological characteristics of water, usually in respect to its suitability for a particular purpose.

Nutrients

Any food, chemical element or compound an organism requires to live, grow, or reproduce. are reaching such proportions, stimulating plant and algal growth in our waters, sufficient to deplete supplies of dissolved oxygen Measures the amount of gaseous oxygen dissolved in an aqueous solution. Oxygen gets into water by diffusion from the surrounding air, by aeration (rapid movement), and as a waste product of photosynthesis. necessary for livelihood of plant and wildlife populations. When waters have too high a concentration of nutrients then it is said to be "eutrophic". Having an excessive supply of nutrients, mostly in the form of nitrates and phosphates. Essentially, when fertilizers are applied to plants in volumes exceeding those absorbed by plant growth, the surplus runs off the land or permeates into ground waters.

One may justly question why *Lyngbya* is so obvious now and was not so obvious long ago, nor even a few decades ago. The principal variant between then and now, is simply the rate at which the water flowed then as opposed to now. There is plenty of evidence illustrating that *Lyngbya* and other algal species do not thrive in faster flowing streams or in better aerated waters.

Think about it. Flow rates were unimpeded many years ago, water levels in the aquifer. Porous, water-bearing layers of sand, gravel, or rock were not reduced, and *Lyngbya* was unable to take hold, except near the edges of a water body where flow rates were slower. Since then much has happened to reduce flow rates so that the algae can take hold more easily. Only a relatively minute difference in the pressure equilibrium is needed to affect the flow rate of the waters across the aquifer and down river. Think how the aquifer pressure reduces, and rates of flow from the springs are lowered, as the many homes, farms, golf courses, businesses, government and industrial buildings take up fresh water. Water that is relatively free of salts. from the aquifer for a whole variety of good purposes, allowing surpluses to evaporate from the increases in impervious. Not allowing fluid to pass through areas. Flow rates have been further reduced by more than a hundred canals cut to allow "waterfront" to so many homes in and around the City of Crystal River, with each canal subject to inward tidal. Relating to or affected by tides. flow and counter flow twice a day. Think of the multitude of boat docks and moored craft which further slow the water flows.

The amount of a component in a given area or volume. of nutrientsAny food, chemical element or compound an organism requires to live, grow, or reproduce. than Crystal River/Kings Bay, but by flowing at a faster rate it has fewer algae and better water clarity.

ALGAL BLOOM EVENT

Algae now seriously impair our local water qualityA term used to describe the chemical, physical, and biological characteristics of water, usually in respect to its suitability for a particular purpose. A rapid growth of microscopic algae or cyanobacteria in water often resulting in a coloured scum on the surface. occurred in the southern part of Kings Bay.

THE SPECIES

A taxonomic category subordinate to a genus (or subgenus) and superior to a subspecies or variety, composed of individuals possessing common characters distinguishing them from other categories of individuals of the same taxonomic level. In taxonomic nomenclature, species are designated by the genus name followed by a Latin or Latinized adjective or noun. of algae in that event has been identified as *Chaetomorpha*. On that Saturday, a very low tide revealed dense algal blooms completely covering an area of some ten acres and more. A taxonomic category subordinate to a genus (or subgenus) and superior to a subspecies or variety, composed of individuals possessing common characters distinguishing them from other categories of individuals of the same taxonomic level. In taxonomic nomenclature, species are designated by the genus name followed by a Latin or Latinized adjective or noun. Algae were involved, including the easily recognized *Lyngbya* and *Spirogyira* among others. Commendable efforts were immediately made to harvest the bloom, over a period of several weeks before it could detach itself, float to the surface and be spread elsewhere by wind and water movements. However, within a couple of weeks of being "cleared", evidence of further bloom throughout the area could be seen. We have a really serious problem on our hands, to combat the spread of algae. *Lyngbya*, particularly, is very aggressive – it will bloom to spread and cover one hundred square meters a minute.

ALGAE (*LYNGBYA*) LIFE CYCLE

Green pigment found in plants and some bacteria used to capture the energy in light through photosynthesis and then they bloom. As it grows, clusters of bacteria Simple single celled prokaryotic organisms. Many different species of bacteria exist. Some species of bacteria can be pathogenic causing disease in larger more complex organisms. Many species of bacteria play a major role in the cycling of nutrients in ecosystems through aerobic and anaerobic decomposition. Finally, some species form symbiotic relationships with more complex organisms and help these life forms survive in the environment by fixing atmospheric nitrogen, associate together in a protective sheath to form a filament structure. New filaments form from clusters of bacteria which emerge enclosed in a membrane from the end of the sheath. As they grow using energy from sunlight the filaments interweave to form mats. Nutrients are absorbed through the sheath walls. Within the bacteria gas vesicles form to regulate buoyancy of the mat. As buoyancy builds up a layer of mat rises, eventually to tear away from the bottom mat and float to the surface. The filaments from the bottom mat can become quite lengthy and smother and kill other aquatic plant material they encounter. The bottom mats continue to grow after yielding layers to the surface. The surface mats travel with wind and water movement to new locations. Counter currents cause the floating mats to clump together into dense areas, sometimes exceeding an acre in size. Floating mats exclude sunlight and otherwise deprive plants and wildlife of habitat. The place or set of environmental conditions in which a particular organism lives.. Prolonged hard rains on the surface matsdisturb buoyancy of the gas vesicles so that the floating mats sink to the bottom. There, in a new place, they can grow and bloom as new mature bottom mats. Some mats become non viable as they are deprived of nutrients Any food, chemical element or compound an organism requires to live, grow, or reproduce or sunshine, or the water temperature falls as cooler weather comes.

As the sun drifts south later in the year and loses sufficient strength for the algae to photosynthesize and grow, the *Lyngbya* goes to sleep as a dark wispy mass on the bottom. What we see as brown or black surface agglomerations are really non viable surface mats in decay. In spring time, as the sun advances to the north, more sunshine energy allows the *Lyngbya* to grow again, and bloom quite dramatically. Lyngbya is more robust than most other algae, having a filament sheath which measures more than sixty microns across. Four toxins have been identified to characterize Lyngbya, and set it apart as a threat, although not all occurrences of the same Lyngbya species have all the toxins, indicating that the presence of toxins may correlate with specific nutrient content. Two of the dermal toxins have been identified in Lyngbya specimens taken from Crystal River/Kings Bay.

Bacteria that have the ability to photosynthesize like *Lyngbya*, is that initial observations may be relatively sparse, only for a sudden bloom to be overwhelming. The slow development takes place largely out of sight below the surface. Too small for human eyesight without employing a microscope growth can then spawn a bloom over several acres in a matter of hours, or at best a few days, is warning indeed of the true nature and maturity of the threat posed to our local Outstanding Florida Water. It has been colonized.

Glucose, and oxygen which is released to the atmosphere. All of the sugar produced in the photosynthetic cells of plants and other organisms is derived from the initial chemical combining of carbon dioxide and water with sunlight. This chemical reaction is catalyzed by chlorophyll acting in concert with other pigment, lipid, sugars, protein, and nucleic acid molecules. Sugars created in photosynthesis can be later converted by the plant to starch for storage, or it can be combined with other sugar molecules to form specialized carbohydrates such as cellulose, or it can be combined with other nutrients such as nitrogen, phosphorus, and sulfur, to build complex molecules such as proteins and nucleic acids. Also see chemosynthesis. It is said that photosynthesis gives rise to three quarters of the world supply of oxygen that we breathe. the vesicles needed for a section to rise could occupy a number of growth cycles.

Vidual long spiral form filaments rather than layers of a bottom mat. It is thought that the spiral may contain gas sufficient for it to float. On the surface, mats form and are moved about as the Lyngbya does. Spirogyra is not a bacteria, individual cells each joined to each other have a controlling nucleus.

Conclusion

We cannot say that the threat to Crystal River/Kings Bay from algae is under control, nor are the circumstances which have given rise to the algae infestation likely to change any time soon. Design enhancements to the harvesters allow *Lyngby*– to be lifted from the bottom, but along with whatever else is growing there above ground. In order to control *Lyngbya* before its penetration becomes so extensive as to disrupt and distort the natural balance so much, that species. A taxonomic category subordinate to a genus (or subgenus) and superior to a subspecies or variety, composed of individuals possessing common characters distinguishing them from other categories of individuals of the same taxonomic level. In taxonomic nomenclature, species are designated by the genus name followed by a Latin or Latinized adjective or noun, which depend for their health upon natural habitat. The place or set of environmental conditions in which a particular organism lives no longer exist for us, nor for our economically significant visitors, to enjoy.

An attractive option would be to contain algal blooms from spreading in spring and summer time, and progressively reduce them by rendering them non-viable and less able to bloom in spring and summer by attacking them while asleep during the wintertime. A further strategy would be to contol the limiting nutrients. Any food, chemical element or compound an organism requires to live, grow, or reproduce. and deprive the algae of their ood source for growth. As either or both means succed, we must be ready to repopulate denuded areas with native aquatic vegetation, to prevent rogue organisms from taking over the vacated areas. This would necessarily involve a systematic process of continually improving management policies and practices, learning from the outcomes of previous policies and practices, also known as Adaptive Management. A management process involving a defined start position and a defined objective position, whereby the progress toward the objective can be measured, in order that judgments may be made as to degree of achievement reached and the process continued in the same or a revised form, or abandoned.

But first, regarding this algae problem, a control mechanism has to be proved effective and safe to apply.

SPIROGYRA

Spirogyra is a genus of filamentous green algae of the order Zygnematales, named for the helical or spiral arrangement of the chloroplasts that is diagnostic of the genus. It is commonly found in freshwater areas, and there are more than 400 species of *Spirogyra* in the world *pirogyra* measures approximately 10 to 100μm in width and may stretch centimeters long.

General Characteristics

Spirogyra is unbranched with cylindrical cells connected end to end in long green filaments. The cell wall has two layers: the outer wall is composed of cellulose while the inner wall is of pectin. The cytoplasm forms a thin lining between the cell wall and the large vacuole it surrounds. Chloroplasts are embedded in the peripheral cytoplasm; their numbers are variable (as few as one). The chloroplasts are ribbon shaped, serrated or scalloped, and spirally arranged, resulting in the prominent and characteristic green spiral on each filament. Each chloroplast contains several pyrenoids, centers for the production of starches, appearing as small round bodies.

Spirogyra is very common in relatively clean eutrophic water, developing slimy filamentous green masses. In spring *Spirogyra* grows under water, but when there is enough sunlight and warmth they produce large amounts of oxygen, adhering as bubbles between the tangled filaments. The filamentous masses come to the surface and become visible as slimy green mats. *Mougeotia* and *Zygnema* are often found tangled together with *Spirogyra*.

REPRODUCTION

Spirogyra can reproduce both asexually and sexually. In asexual reproduction, fragmentation takes place, and *Spirogyra* simply undergoes intercalary mitosis to form new filaments.

Sexual Reproduction is of two types:

— Scalariform conjugation requires association of two different filaments lined side by side either partially or throughout their length. One cell each from opposite lined filaments emits tubular protuberances known as conjugation tubes, which elongate and fuse, to make a passage called the conjugation canal. The cytoplasm of the cell acting as the male travels through this tube and fuses with the female cytoplasm, and the gametes fuse to form a zygospore.

— In lateral conjugation, gametes are formed in a single filament. Two adjoining cells near the common transverse wall give out protuberances known as conjugation tubes, which further form the conjugation canal upon contact. The male cytoplasm migrates through the conjugation canal, fusing with the female. The rest of the process proceeds as in scalariform conjugation.

The essential difference is that scalariform conjugation occurs between two filaments and lateral conjugation occurs between two adjacent cells on the same filament.

Chapter–12

MARINE ALGAE

INTRODUCTION

Marine algae, or seaweeds, are the oldest members of the plant kingdom, extending back many hundreds of millions of years. They have little tissue differentiation, no true vascular tissue, no roots, stems, or leaves, and no flowers. Algae range in size from microscopic individual cells to huge plants more than 100 feet long. Though the flora is continuous along the coast, an abrupt change in overall species composition occurs at Point Conception, where nutrient-rich northern currents meet warmer southern ones.

Zonation patterns within algal assemblages are dictated by tidal exposure and wave impact, as well as by species interactions such as grazing by invertebrates, and by competition for space and light. The sea palm, *Postelsia palmaeformis*, for example, finds refuge on wave-pounded rocks where predators, such as sea urchins, are unable to follow. The sea palm's tough, cartilaginous stipe and holdfast (analogous to a stem and root in vascular plants) absorb wave shock, and the thick slippery cell walls reduce desiccation. In contrast, calmer waters support more delicate membranous and leaflike seaweeds such as *Botryoglossum*, *Porphyra*, and *Plocamium* species.

Common algae of the high intertidal include Turkish towel, *Mastocarpus papillatus*, and rockweed, *Fucus distichus*.

Common in the middle and lower intertidal are sea lettue, *Ulva* sp.; feather boa kelp, *Egregia menziesii*; dead man's fingers, *Codium fragile*; and the iridescent, rubbery *Iridaea* species.

Kelp forests, found in subtidal waters as deep as 100-200 feet, are a unique ecosystem largely restricted to the west coast of the Americas, and best represented in California's coastal waters. Giant Kelp, *Macrocystis pyrifera*, and bull kelp, *Nereocystis luetkeana*, are the largest non-vascular plants known. Their blades are harvested for industrially valuable gels, called alginates. The multi-layered canopy of kelp fronds provides a complex aquatic habitat for thousands of fish and invertebrates.

In addition to seaweeds, marine flowering plants called seagrasses are also present in some intertidal assemblages. These plants can dominiate mudflats and exposed rocks, in some cases virtually excluding the algae. Eelgrass, *Zostera marina*, is common in quiet waters with a sandy or muddy substrate, whereas surfgrass, *Phyllospadix* sp., is common along exposed rocky shores, often in high-energy surf zones.

Algae are vitally important to marine ecosystems, and most species of algae are not harmful. However, some species of microscopic marine algae produce powerful toxins that can harm animals and humans. People who come in contact with these toxins by eating fish that feed on harmful algae, swimming among harmful algae, or breathing air that contains toxins from harmful algae, may experience neurologic symptoms (such as tingling fingers and toes), respiratory, or gastrointestinal symptoms.

CDC Focus Areas

Ciguatera: Ciguatera fish poisoning (or ciguatera) is an illness humans can get by eating fish that contain toxins produced by the microscopic marine algae *Gambierdiscus toxicus*. People who have ciguatera may experience nausea, vomiting, and neurologic symptoms such as tingling fingers or toes. They also may find that cold things feel hot and hot things feel cold. Ciguatera has no cure. Symptoms usually go away in days or weeks but can last for years. People who have ciguatera can receive treatment for their symptoms.

Investigators at CDC, the Food and Drug Administration, the University of Miami, and Jackson Memorial Hospital in Miami are collaborating to develop a test that could be used to identify whether someone has been exposed to the toxins that cause ciguatera.

Red tides occur when the microscopic marine algae *Karenia brevis* (*K. brevis*) grow quickly, creating blooms that make the ocean appear red or brown. *K. brevis* produces potent toxins called brevetoxins, which have killed millions of fish and other marine organisms. Red tides have damaged the fishing industry, shoreline quality, and local economies in states such as Texas and Florida. In addition to killing fish outright, brevetoxins can become concentrated in the tissues of shellfish that feed on *K. brevis*. People who eat these shellfish may suffer from neurotoxic shellfish poisoning, a food poisoning that can cause neurologic symptoms (such as tingling fingers or toes) and severe gastrointestinal symptoms.

The human health effects associated with eating shellfish that contain high concentrations of brevetoxins are well documented. However, scientists know little about how other types of environmental exposures to brevetoxin — such as breathing the air near red tides or swimming in red tides — may affect humans. For more information about red tide, and to learn what CDC is doing to study environmental exposures to brevetoxins, visit CDC's red tide.

There are many examples of plants, bacteria and algae that have formed intimate symbiotic associations or "marriages" with each other. Divorce is practically nonexistent in these marriages, and separations may result in the death of one or both partners. In some cases the relationship is decidedly one-sided, with only one partner actually benefiting. These relationships are often termed parasitic, especially when the non-benefiting partner is actually harmed by the relationship. In other marriages the relationship is mutually beneficial. Algae and fungi live together in an association called lichen, and nitrogen-fixing bacteria live symbiotically inside the root nodules of legumes. But one of the most fascinating of all plant marriages involves a tiny aquatic water fern (Azolla) and a microscopic filamentous blue-green alga or cyanobacterium (*Anabaena azollae*). They grow together at the surface of quiet streams and ponds throughout tropical and temperate regions of the world.

Ponds along the San Dieguito River (San Diego County, California) are covered with a reddish carpet of Azolla filiculoides during the fall months. Photo also shows clump of cattails (*Typha latifolia*) and naturalized Australian *Eucalyptus camaldulensis*. The Eucalyptus trees were introduced into California near the turn of the century for fast-growing hardwood lumber for railroad ties, but proved inadequate because the spikes would not hold in the badly checked wood. Now these trees have literally taken over parts of San Diego County.

AZOLLA PLANTS

Virtually any sample of Azolla examined under a microscope will have filaments of Anabaena living within ovoid cavities inside the leaves. Like nitrogen-fixing bacteria living inside the root nodules of legumes, the relationship appears to be mutually beneficial. Since Azolla is easy to maintain in aquarium cultures, it is an excellent source of prokaryotic cells and heterocysts for general biology laboratory exercises on cell structure and function. It also has an interesting heterosporous life cycle and can readily be adapted to laboratory exercises on symbiosis. In addition, this little fern and its algal partner provide an important contribution toward the production of rice for a hungry world.

It has been estimated that there are at least 10,000 different species of ferns in the world, from large tree ferns of tropical rain forests to small rock ferns of desert canyons and alpine crevices. Fossil evidence indicates that many additional species of ferns flourished on earth during the Carboniferous period, some 300 million years ago. But of all the great diversity of ferns, relatively few kinds have colonized the water. Azolla belongs to the Salvinia Family (Salviniaceae), although some authorities now place it in the monotypic family, Azollaceae. Six species are distributed worldwide, three of which occur in the United States: *A. filiculoides, A. mexicana* and *A. caroliniana*.

Individual Azolla plants have slender, branched stems with minute, overlapping scalelike leaves only one millimeter long. Each plant resembles a little floating moss with slender, pendulous roots on its underside. The plants tend to clump together and often form compact mats on the water surface. Azolla is sometimes called "duckweed fern" and commonly grows with one or more species of duckweeds (Lemnaceae), including Lemna, Spirodela, Wolffia and Wolffiella. When growing in full sunlight, particularly in late summer and fall, Azolla may produce reddish anthocyanin in the leaves, in contrast with the bright green carpets of duckweed and filamentous green algae.

WATER FERN PLANTS

Several water fern plants (Azolla filiculoides) floating on the water surface. The plant in upper right with oval fronds is a duckweed (*Lemna minuta*). The minute plants which resemble tiny green bubbles are *Wolffia borealis* and another interesting species *W. columbiana*, two of the world's smallest flowering plants.

The stems of Azolla pull apart readily and the plants reproduce by fragmentation at an astonishing rate. In fact, some species can double their biomass in three days under optimal environmental conditions. Entire ponds and muddy banks may be completely covered by a carpet of velvety green or pink. In addition to providing food for various water fowl, Azolla also provides the habitat and food for numerous kinds of freshwater insects, worms, snails and crustaceans.

A related African water fern (*Salvinia rotundifolia*), also listed as S. auriculata in some floras, is mentioned in the *Guinness Book of World Records* (1985 UK Edition) as the "most intransigent weed." This mat-forming aquatic fern was detected on the filling of Kariba Lake in May 1959. Within 11 months it covered an area of 77 square miles (199 km2), and by 1963 it covered 387 square miles (1002 km^2).

The water fern (*Salvinia rotundifolia*), a ubiquitous floating fern in quiet waters of streams and ponds throughout tropical America, Africa and Florida.

SEXUAL REPRODUCTION IN AZOLLA

Not only are Azolla species prolific vegetative reproducers, but they also have a very interesting and uniquely specialized sexual cycle. Like all ferns, Azolla produces spores; however, unlike most ferns, Azolla produces two kinds of spores. If you carefully examine *Azolla filiculoides* during the summer months you can easily find numerous spherical structures called sporocarps on the undersides of the branches. The sporocarp of **Azolla** is homologous to the sorus of other ferns, and the sporocarp wall represents a modified indusium. The male sporocarp is a greenish or reddish case about two millimetres in diameter, and inside are numerous male sporangia which look like the egg mass of an insect or spider inside a transparent case. Male spores (microspores) are extremely small and are produced inside each microsporangium.

Close-up view of *Azolla filiculoides* showing scalelike, overlapping leaves and several globose reproductive structures called sporocarps. The male sporocarp (middle right) contains microsporangia that resemble eggs inside an egg sac. One sporocarp has broken open releasing many spore clusters or massulae. The minute ovoid plants in center are Wolffia borealis, a minute flowering seed plant. Closeup view of Azolla filiculoides showing a small female sporocarp flanked by two larger, globose male sporocarps. These tiny structures are so small that they could easily fit on the head of an ordinary straight pin. One very curious thing about microspores is that they tend to stick together in little clumps or masses called massulae. In the American species (*Subgenus euazolla*), each spore mass (massula) is covered with minute hairs (barbed at the tips) called *glochidia*. Under high magnification the massulae look like strange space satellites with radiating antennae. The female sporocarps are much smaller, and contain a single sporangium and a single functional spore. Since an individual female spore is considerably larger than a male spore, it is termed a megaspore.

Highly magnified view (400X) of one spore mass (massula) of *Azolla filiculoides* with unique barbed projections called glochidia. In a related species of water fern (*A. mexicana*), the glochidia are septate with several distinct partitions. Although rarely seen by the casual observer, the spores of most ferns develop into a fleshy, heart-shaped structure called a prothallus or gametophyte, which produces the actual sex organs (the female archegonium and male antheridium). Azolla has microscopic male and female gametophytes that develop inside the male and female spores. The female gametophyte protrudes from the megaspore and bears one to several archegonia, each containing a single egg. The microspore forms a male gametophyte with a single antheridium which produces eight swimming sperm. The barbed glochidia on the male spore clusters presumably cause them to cling to the female megaspores, thus facilitating fertilization.

According to some references, the universal occurrence of Anabaena azollae inside the leaves of Azolla suggests that reproduction of this water fern may be chiefly vegetative; however, it has been shown that this cyanobacterium and fern may develop in synchrony. Short filaments of Anabaena, called hormogonia, often survive under the "indusium cap," on top of the germinating megaspore. Hormogonia may be entrapped by the embryo Azolla plant during differentiation of the shoot apex and dorsal lobe primordia of the first leaves. It is fascinating to speculate on just how and when these two diverse organisms formed such an intimate association. Although filamentous cyanobacteria (with cells resembling heterocysts) date back more than two billion years, fossil Azolla plants are known only from late Cretaceous deposits less than 80 million years ago.

ANABAENA AND NITROGEN FIXATION

Close examination of an Azolla leaf reveals that it consists of a thick, greenish (or reddish) dorsal (upper) lobe and a thinner, translucent ventral (lower) lobe emersed in the water. It is the upper lobe that has an ovoid central cavity, the "living quarters" for filaments of Anabaena. Probably the easiest way to observe Anabaena is to remove a dorsal leaf lobe and place it on a clean slide with a drop of water. Then apply a cover slip with sufficient pressure to mash the leaf fragment. Under 400X magnification the filaments of Anabaena with larger, oval heterocysts should be visible around the crushed fern leaf. The thick-walled heterocysts often appear more transparent and have distinctive "polar nodules" at each end of the cell. According to Dr. C. P. Wolk at Michigan State University Plant Research Laboratory, the "polar nodules" may be the same composition as cyanophycin granules (co-polymer of arginine and aspartic acid). Cyanophycin granules occur in many cyanobacteria and may serve as a nitrogen storage product.

Modern-day filamentous cyanobacteria (*Anabaena azollae*) from cavities within the leaves of the ubiquitous water fern (*Azolla filiculoides*). The larger, oval cells are heterocysts (red arrow), the site of nitrogen-fixation where atmospheric nitrogen (N_2) is converted into ammonia (NH_3). Polar nodules are visible in some of the heterocysts. The water fern benefits from its bacterial partner by an "in house" supply of usable nitrogen. The cellular structure of these bacteria has changed very little in the past one billion years. Nitrogen fixation is a remarkable prokaryotic skill in which inert atmospheric nitrogen gas (N_2) is combined with hydrogen to form ammonia (NH_3). This vital process along with nitrification (formation of nitrites and nitrates) and ammonification (formation of ammonia from protein decay) make nitrogen available to autotrophic plants and ultimately to all members of the ecosystem. Although Azolla can absorb nitrates from the water, it can also absorb ammonia secreted by Anabaena within the leaf cavities. Recent studies have shown that the actual site of nitrogen fixation occurs within the thick-walled heterocysts. As the heterocyst matures, the photosynthetic membranes (thylakoid membranes) become contorted or reticulate compared to regular photosynthetic cells of Anabaena, and they become non-photosynthetic (and do not produce oxygen). This fact is especially noteworthy because nitrogen fixation requires the essential enzyme nitrogenase, and the activity of nitrogenase is greatly inhibited by the presence of oxygen.

AZOLLA AND RICE PRODUCTIVITY

Rice is the single most important source of food for people and Azolla plays a very important role in rice production. For centuries Azolla and its nitrogen-fixing partner, Anabaena, have been used as "green manure" in China and other Asian countries to fertilize rice paddies and increase production. Some authorities believe the use of Azolla enabled the Vietnamese to survive the effects of the American blockade when imported fertilizers did not reach North Vietnam during the war. According to Wilson Clark (Science 80: Scpt./Oct. 1980), the People's Republic of China has 3.2 million acres of rice paddies planted with Azolla. This provides at least 100,000 tons of nitrogen fertilizer per year worth more than $50 million annually. Extensive propagation research is being conducted in China to produce new varieties of Azolla that will flourish under different climatic and seasonal conditions. According to some reports, Azolla can increase rice yields as much as 158 percent per year. Rice can be grown year after year, several crops a year, with little or no decline in productivity; hence no rotation of crops is necessary.

In addition to nitrogen fixation, Azolla has a number of other uses. Several California aquafarms grow Azolla in large vats of circulating fresh water. Apparently fish and shrimp relish the Azolla. In fact, Azolla was grown for fish food and water purification at the Biospere II project in Arizona (a 2.5 acre glass enclosure simulating an outer space greenhouse). Fresh Azolla and duckweed (Wolffia) can also be used in salads and sandwiches, just as alfalfa and bean sprouts are used. Dried, powdered Wolffia and Azolla make a nutritious, high protein powder similar to the popular alga (cyanobacterium) Spirulina that is sold in natural food stores. Azolla has also proved useful in the biological control of mosquitos. The mosquito larvae are unable to come up for air because of the dense layer of Azolla on the water surface. Azolla grows very quickly in ponds and buckets, and in makes an excellent fertilizer (green manure) and garden mulch.

A male tree frog (*Hyla regilla*) floating in a pond of *Azolla filiculoides*. On a warm summer night, the amazing chorus of dozens of these small frogs can be almost deafening. Azolla is certainly a valuable laboratory plant that will thrive with very little care. In fact, at the WAYNE'S WORD headquarters, the hippo staff has maintained buckets of Azolla in ordinary tap water, although it may exhibit seasonal fluctuations in density and vigor. Depending on the sophistication of viewing equipment and level of study, this plant can fascinate biology students from junior high school to college. But its value goes far beyond the classroom. The use of Azolla may be an important factor in the world's future food needs and may play an important role in reducing the world's reliance on fossil fuel-based fertilizers. The significance of its symbiotic relationship with Anabaena is astounding when one considers that millions of lives depend on these two organisms.

SYMBIOSES

Symbioses are intimate associations involving two or more ecies. Symbioses areincredibly diverse. Familiar examples include corals (associations between cnidarians[animals] and dinoflagellates [unicellular algae]), lichens (fungi and green algae or cyanobacteria), and malaria (involving the apicomplexan, *Plasmodium*, and its mosquitoand human hosts). The establishment of symbioses has shaped the evolution of individual clades and entire ecosystems. For example, it has been suggested that the evolution of mycorrhizal symbioses (involving plants and fungi) as a key innovation that allowed plants to colonize the land and establish terrestrial ecosystems.

Symbiosis is a major source of evolutionary and ecological novelty. It is therefore of general interest to understand the processes that establish, maintain, and disrupt symbioses. Symbioses are mentioned in many biology courses, but one rarely encounters a course that takes symbiosis as its central focus. This course will survey the diversity, evolution, and natural history of symbioses through articles in the primaryliterature. These research articles will focus on specific questions concerning individual symbioses. By the end of the semester, we hope that we will be able to address the following general questions concerning the natural history and evolution of symbioses:

1. Are symbioses stable endpoints in evolution?
2. Do symbioses begin as antagonistic interactions and tend to evolve toward mutualism?
3. Does symbiosis promote or retard speciation?
4. Do some groups of organisms enter into symbioses more easilythan others?

This course is also about the process of communicating in science. Scientific communication takes many forms, but principally it includes writing, presentation, discussion, and reviewing the work of others. This course will provide practical experience in all of these activities and is intended to help you learn how to function asindependent scientists.

Chapter–13

Algal Blooms in Fresh Water

INTRODUCTION

Aquatic ecologists are concerned with blooms (very high cell densities) of algae in reservoirs, lakes, and streams because their occurrence can have ecological, aesthetic, and human health impacts. In waterbodies used for water supply, algal blooms can cause physical problems (e.g., clogging screens) or can cause taste and odor problems in waters used for drinking. Blooms involving toxin-producing species can pose serious threats to animals and humans.

The term "algae" is generally used to refer to a wide variety of different and dissimilar photosynthetic organisms, generally microscopic. Depending on the species, algae can inhabit fresh or salt water.

In modern taxonomic systems, algae are usually assigned to one of six divisions The misnamed blue-green algae are often grouped with algae because of the chloroplasts contained within the cells. However, these organisms are actually photosynthetic bacteria assigned to the group cyanobacteria.

Fresh-water algae, also called phytoplankton, vary in shape and color, and are found in a large range of habitats, such as ponds, lakes, reservoirs, and streams. They are a natural and essential part of the ecosystem. In these habitats, the phytoplankton are the base of the aquatic food chain. Small fresh-water crustaceans and other small animals consume the phytoplankton and in turn are consumed by larger animals.

BLOOM OCCURRENCES AND IMPACT

Under certain conditions, several species of true algae as well as the cyanobacteria are capable of causing various nuisance effects in fresh water, such as excessive accumulations of foams, scums, and discoloration of the water. When the numbers of algae in a lake or a river increase explosively, an algal "bloom" is the result. Lakes, ponds, and slow-moving rivers are most susceptible to blooms.

Algal blooms are natural occurrences, and may occur with regularity (e.g., every summer), depending on weather and water conditions. The likelihood of a bloom depends on local conditions and characteristics of the particular body of water. Blooms generally occur where there are high levels of nutrients present, together with the occurrence of warm, sunny, calm conditions. However, human activity often can trigger or accelerate algal blooms. Natural sources of nutrients such as phosphorus or nitrogen compounds can be supplemented by a variety of human activities. For example, in rural areas, agricultural runoff from fields can wash fertilizers into the water. In urban areas, nutrient sources can include treated wastewaters from septic systems and sewage treatment plants, and urban stormwater runoff that carries nonpoint-source pollutants such as lawn fertilizers.

An algal bloom contributes to the natural "aging" process of a lake, and in some lakes can provide important benefits by boosting primary productivity. But in other cases, recurrent or severe blooms can cause dissolved oxygen depletion as the large numbers of dead algae decay. In highly eutrophic (enriched) lakes, algal blooms may lead to **anoxia** and fish kills during the summer. In terms of human values, the odors and unattractive appearance of algal blooms can detract from the recreational value of reservoirs, lakes, and streams. Repeated blooms may cause property values of lakeside or riverside tracts to decline.

TOXIC BLOOMS

Some algae produce toxic chemicals that pose a threat to fish, other aquatic organisms, wild and domestic animals, and humans. The toxins are released into the water when the algae die and decay.

The most common and visible nuisance algae in fresh water, and the species that are often toxic, are the cyanobacteria. A cyanobacterial bloom will form on the surface and can accumulate downwind, forming a thick scum that sometimes resembles paint floating on the water. Because these mats are blown close to shore, humans and wild and domestic animals can come into contact with the unsightly material.

Blooms of toxic species of algae and cyanobacteria can flood the water environment with the biotoxin they produce. When toxic, blooms can cause human illnesses such as gastroenteritis (if the toxin is ingested) and lung irritations (if the toxin becomes aerosolized and hence airborne). Other cyanobacterial toxins are less drastic, and cause skin irritation to people who swim through an algal bloom. Toxicity can sometimes cause severe illness and death to animals that consume the biotoxin-containing water.

Cyanobacterial toxins are known to affect bean photosynthesis when they are present in irrigation water. The toxins also can modify zooplankton communities, reduce growth of trout, and interfere with development of fish and amphibians. In some cases, toxins can be bioconcentrated by fresh-water clams.

Some algal blooms in fresh water may only be a nuisance, but others can deplete dissolved oxygen in the water or generate biotoxins that are harmful to birds, fish, and other animals. This Canada goose swims among a floating layer of heavy, but probably harmless, algal growth.

Microcystins comprise the most common group of about fifty cyanobacterial toxins. Among these toxins are ones that, if ingested in sufficient quantity, can harm the liver (hepatotoxins) or nervous system (neurotoxins). Microcystins can persist in water because they are stable in both hot and cold water. Even boiling the water, which makes the water safe from harmful bacteria, will not destroy microcystins. As a result of this threat, the Canadian government implemented a recommended water-quality guideline of 1.5 µg per liter of microsystin-LR (the most common hepatotoxin), and other countries will likely follow suit. In Canada as well as the United States, there are few reports of injury and no reports of human deaths resulting from microcystins in drinking water, in large part because surface-water sources of drinking water (e.g., reservoirs, lakes, and rivers) must undergo filtration and chlorination at water utilities prior to being distributed to customers. (Cyanobacterial toxins can be removed from water only by activated charcoal filters and chlorination.)

CONTROL CONSIDERATIONS

Repeated episodes of algal blooms can be an indication that a river or lake is being contaminated, or that other aspects of a lake's ecology are out of balance. While cyanobacterial blooms receive the most public and scientific attention, the excessive growth of other algae and other aquatic plants also can cause significant degradation of a lake or pond, particularly in waters receiving sewage or agricultural runoff. Aquatic biologists and other water-quality specialists often are called to identify the causes and recommend management steps to reduce or control the problem.

However, prevention of a problem is always better than trying to fix the problem after it happens. Controlling agricultural, urban, and stormwater runoff; properly maintaining septic systems; and properly managing residential applications of fertilizers are probably the most effective measures that can be taken to help prevent human-induced fresh-water algal blooms.

The Five Kingdoms

Scientists use a system called taxonomy to organize all the biological organisms in the world. Organisms are put into various classification groups according to the distinguishing properties they share. These groups are (from highest to lowest) kingdom, phylum, class, order, family, genus, and species.

Although there are several different kingdom classifications in use, it is now generally accepted that all biological organisms can initially be placed into one of five kingdoms: monera, protists, fungi, plants, and animals.

Prokaryotes (Cells That Have No Distinct Nuclei)

- *Monera:* Includes aquatic bacteria and blue-green algae, more properly called cyanobacteria. Monerans, though microscopic, are the most dominant organisms on Earth. They have existed for about 3.5 billion years.

Eukaryotes (Cells Have Distinct Nuclei)

- *Protista:* Includes plant-like and animal-like primitive organisms, such as algae and protozoa. Organisms are generally unicellular.
- *Fungi:* Includes a large group of parasitic and saprophytic species. Some are parasitic on animals, including humans (ringworm, or athlete's foot). Others are parasitic on plants and include rusts and mildews. Fungi are, along with the bacteria, important decomposers of dead organic matter.
- *Plants:* Make their food by the process of photosynthesis.
- *Animals:* Ingest their food and digest it internally in specialized body cavities.

ECOLOGY, FRESH-WATER

Streams, lakes, and wetlands differ profoundly from one another in the conditions they provide as habitats for biological communities. Fundamental characteristics of standing water (a lentic system) or flowing water (a lotic system), the dynamics of its interaction with adjacent land and vegetation, and seasonal fluctuations in water conditions determine characteristic biological assemblages.

The most apparent feature of streams is flowing water. The geomorphology and topography through which streams flow, their channel steepness, the variability in stream-bottom substrates, and the availability of woody debris are among the natural controls that influence flow patterns. The diversity in physical structure and flow among riffles, pools, and glides is reflected in equally diverse aquatic communities.

A flowing-water habitat is known as a lotic ecosystem. A lentic system, such as a lake, is predominantly non-flowing.

In flat valley floodplains, streams naturally meander, or flow in a winding path. In these systems, flooding is an important natural process: nutrient-rich sediments are deposited onto the floodplain, and young fish are protected in quieter areas away from high flow. Terrestrial (land-based) invertebrates become important sources of food during these times. Under the stream and beneath floodplain soils, water flows in the subsurface in the hyporheic zone. Waters are cooled, nutrients are exchanged, and invertebrates are specially adapted to live in this subterranean habitat.

A river basin consists of its entire upstream network: its watershed is the landscape, including the stream network and the land it traverses. Riparian (streamside) zones make up the transition between stream and terrestrial environment, encompassing the area where vegetation and streams interact. Riparian plants provide shade, cool the stream, retain soils, and filter nutrients. Fallen leaves decompose and provide nutrients for stream microorganisms and invertebrates.

Changing habitats along the longitudinal stream gradient, from its headwaters to downstream reaches, can be thought of as a stream continuum. Energy possessed by the stream and aquatic communities tend to follow this continuum in predictable ways. Leafy inputs are most important in headwaters, whereas algae are more important as a food base in wider, more sunlit reaches. Nutrients released from streambanks and within the stream are taken up by algae in a downstream spiral of repeated uptake and release.

Macro-invertebrates in the stream continuum are classified into functional feeding groups, depending on how they obtain food. In headwaters, shredding invertebrates, dependent on decomposing leaves, are common. As streams open up, scraping macroinvertebrates that scoop algae off rocks are more abundant. Collector invertebrates either gather or filter out small edible particles everywhere, but in wide, deep rivers they dominate with plankton. As adults many stream invertebrates move into the terrestrial environment.

In stream food webs, primary production (the foundation of food chains) comes from riparian inputs and algae, and most invertebrates are herbivores or detritivores. Top predators are primarily fish but also include birds, amphibians, and humans. Invertebrate predators are relatively few. Fish move easily between habitats, and many feed on both stream and riparian invertebrates. A few birds spear large invertebrates that avoid fish predation. Other birds and bats capture adult aquatic insects that emerge from the stream.

Lakes

Lakes are characterized by basin shape, volume, and depth. Shallow edges where rooted vegetation persists are nursery grounds for fish and diverse invertebrates. The open-water (pelagic) zone is classified into layers, depending on the degree of mixing that occurs. Water changes density with temperature, causing lakes to become layered, or stratified, into temperature zones. Consequently, significant energy is needed to mix a lake's thermal layers. Deep lakes are stratified into a warmer, upper layer that mixes readily (epilimnion); a middle layer of quickly changing temperatures (metalimnion); and a deep, cool bottom layer that mixes infrequently during the year (hypolimnion). Wind easily mixes shallow lakes, so these layers either do not persist or do not develop.

Phytoplankton, predominantly algae, form the base of a lake's food chain. These primary producers fall into five major categories. These major categories also are represented in streams.

A community of microscopic zooplankton lives in a lake's upper layer, feeding on single-celled or small colonial algae. In clear, relatively unproductive lakes, zooplankton consume much of the algae. Very productive lakes are much less clear, with abundant algae in blooms of filamentous forms.

Water chemistry plays an important role in lake dynamics, as nutrients influence algal productivity and higher trophic levels. In deep lakes, the bottom layer has little oxygen when it is not mixing, and few organisms survive there. Similarly, very salty lakes contain only a few highly specialized zooplankton.

When fish occur in lakes, they are the dominant predators. Fish are visual predators and select prey on the basis of size. They consume the largest zooplankton, and remaining zooplankton will be small. A trophic cascade occurs as the top predators (fish) depress the next trophic level (zooplankton), releasing high production in a third level (algae). In fishless lakes, invertebrate predators dominate, and large herbivorous zooplankton escape predation.

Wetlands

Wetlands generally are more shallow than lakes and have standing or flowing water above, near, or below the surface. They vary in size, from puddles to thousands of hectares (a hectare is 10,000 square meters). Because water is not necessarily present year-round, characteristic waterlogged soils or plants help identify their location and extent.

Wetlands often are intermediate habitats, representing a transition between aquatic and terrestrial habitats. Their transitional position makes them key exporters of organic materials and sinks for nutrients. Persistent, shallow standing water forms a permanent wetland. When soils are saturated for only part of the year, temporary or ephemeral wetlands are created.

Wetland communities reflect the duration and depth of standing water. Plants vary specifically in requirements for annual wetness. Similarly, invertebrates differ in intervals required to mature, eventually migrating or becoming dormant in the dry season. Some amphibians use only a seasonal aquatic habitat before migrating; others require continuous wetlands.

Ephemeral plant and invertebrate resources will attract migratory waterfowl, but resident birds and fish require year-round water. Thus, a variety of wetlands support diverse plant and animal communities.

Human Influences

Habitat loss due to human activities has severely affected fresh-water ecosystems. Because of filling and draining activities more than half (53%) of wetlands in the continental United States were lost between 1780 and 1980. Dams, channelization, wood removal, and flow alteration have reduced stream habitat significantly. Profound changes from accidental and intentional introduction of nonnative plant and animal species have altered aquatic communities and destroyed habitats. Acid rain and other consequences of air and water pollution have severely damaged lakes, streams, and wetlands. Cumulative effects of these activities often result in declining native flora and fauna. Although scientists and decisionmakers recognize many alterations caused by humans, they have just begun to develop scientific and political understandings for fresh-water restoration.

ALGAL BLOOMS, HARMFUL

Single-celled algae are almost always present in sea water even if the water looks clear. When high concentrations of certain species of dinoflagellates are present, patches of water look red because these algae contain red pigments—hence the name "red tide." High concentrations of other algae.

The term "red tide" refers to different types of algal blooms, which can be various hues depending on the species and the photosynthetic pigments they contain. Blooms occurring offshore can sometimes float toward the shore, as shown here off the southwestern coast of Molokai, Hawaii. Some blooms have the potential to be harmful to ocean life and humans. may turn sea water orange, yellow, brown, or purple. Red tides have been witnessed for centuries and have been seen all over the world.

Dense concentrations of algae are referred to as blooms, because the algae have multiplied rapidly to become concentrated in high numbers. In a bloom, there could be tens of millions of cells in a liter of sea water. Most blooms are not harmful, but some have the potential to be harmful, whether by virtue of natural biotoxins (poisons) produced by certain species of algae, or by the oxygen-depleting process initiated upon the death and subsequent decay of large concentrations of algae.

Harmful algal blooms, or HABs, cause millions of dollars in damage when there are massive fish kills to be cleaned up, beaches declared offlimits, fisheries and shellfisheries closed to harvesting, and medical treatment provided for people poisoned by marine biotoxins in the seafood they ate. Many scientists believe that harmful algal blooms are becoming more prevalent, but they point out that increased monitoring efforts are detecting more occurrences.

Mechanisms of Harm

How do certain microscopic algae—that is, types of phytoplankton—cause harm to fish, shellfish, marine mammals, seabirds, and people? Basically, there are four ways.

First, the physical presence of so many cells may suffocate fish by clogging or irritating the gills. Second, when the densely concentrated algal cells die off, the decay process, assisted by bacteria, can deplete the water of oxygen, which in turn can lead to the death of oxygen-dependent marine creatures. (Algae, being plants, require nutrients such as nitrogen and phosphorus to grow. When they have used up the nutrients, they tend to die off all at once.) Such oxygen-related impacts are most visible in shallow bays, inlets, or seas.

Third, some algal species produce deadly toxins which directly kill the animals that ingest the poisons. Dinoflagellate toxins have killed mussels,

Fourth, shellfish such as mussels, clams, and oysters feed by filtering particles, including phytoplankton, from sea water. Toxins from certain dinoflagellate or diatom species accumulate in the tissues of shellfish. When people, sea mammals, or seabirds eat the shellfish, they ingest the toxins as well.

There are different kinds of toxins that cause different kinds of symptoms which, in humans, typically are neurological. Some toxins are deadly. The table above shows the categories of human poisoning; the organism associated with the toxicity; and the causative toxin (sometimes among a suite of toxins).

TYPES AND EXAMPLES OF TOXICITY

The algal species that can be toxic in some circumstances are not always toxic. When they are toxic, they cause harm by being eaten by larger organisms. As a toxin is passed up the food chain, it becomes concentrated in larger and larger animals such as fish and shellfish, and eventually is ingested by people who eat seafood containing the toxin. In the following overview, only two examples of the various algal species known to cause toxicity are discussed.

Pfiesteria

An exception to toxin transmission up the food chain is the dinoflagellate Pfiesteria. Instead of being eaten, it does the eating—usually small organisms but also fish. It uses its toxins to make the fish lethargic and to injure the fish's skin.

Toxins can also get into the air and cause harm to people, as happened in 1991 when this peculiar organism was first discovered in a laboratory in North Carolina. Later it was found in connection with fish kills in the Albemarle–Pamlico estuary and other estuarine environments on the U.S. Atlantic and Gulf of Mexico coasts. Blooms of the more common dinoflagellate, *Gymnodinium breve,* when present in nearshore waters, can be picked up by surf and wind and carried to the seashore in the air. The microscopic organisms cause skin and eye irritation in people exposed to this toxic aerosol.

A red-tide dinoflagellate typically has a vegetative stage, in which it multiplies by cell division, and a cyst stage, in which two cells combine to form gametes enclosed in a cyst (a type of covering). The cysts sink to the seafloor until conditions are favorable for a return to the vegetative stage.

The *Pfiesteria* organism has a minimum of twenty-four stages in its life cycle, of which at least four are toxic. The life stages include flagellated cells that swim in the water, amoeboid forms both in the water and in bottom muds, and cysts that rest on the bottom. The different forms vary in size from too small to see with an ordinary microscope, to a speck visible to the naked eye.

Most of the time, the *Pfiesteria* dinoflagellate is a nontoxic predator that feeds on small organisms such as algae, bacteria, and small animals. It becomes toxic when cyst forms detect fish excretions or secretions. Encysted cells emerge and become toxic. They damage the fish with the toxin, then feed on the epidermal tissue, blood, and other substances that leak from sores on the incapacitated fish. When the fish are dead, the cells change to the amoeboid stages and feed on the fish carcass.

Pseudo-nitzschia

Diatoms are single-celled marine or fresh-water algae that have shell-like structures, called frustules, made of silica. Until recently, diatoms were not associated with biotoxin poisonings. But in 1987, an outbreak of domoic acid poisoning was reported in Canada. The domoic acid came from a diatom, *Pseudo-nitzschia*. Domoic acid has caused permanent memory loss and death in humans.

In 1991, examination of the stomach contents of dead seabirds found along the beaches of Monterey Bay, California revealed high levels of domoic acid. The birds had been eating anchovies which had been consuming *Pseudo-nitzschia*. In 1998, sea lion deaths on the California coast also were associated with domoic acid, which entered the food chain via toxigenic diatoms, which were eaten by anchovies that in turn were eaten by the sea lions. To prevent human illness when such toxicity is found, state health departments temporarily close beaches and federal regulatory agencies temporarily close fisheries and shellfisheries along the affected coast.

Human Impacts and Intervention

Although the economic impacts of HAB outbreaks has not been quantified on a national basis, the direct and indirect costs to even a single fishery closure can reach millions of dollars. In addition to loss of revenue to fish and shellfish industries, there are impacts on recreational fishing and tourism and their associated businesses.

Harmful algal blooms also can threaten the aquaculture industry. For example, unpredictable and destructive blooms of the small flagellate *Heterosigma* have threatened the commercial farmed salmon industry in Washington state (USA) and British Columbia (Canada). *Heterosigma* blooms also have destroyed some captive populations of threatened and endangered salmon being raised in net pens before their release to the wild.

All of the U.S. coastal states have developed monitoring programs with regular testing of fish and shellfish from beaches. Officials and volunteers watch the shores for patches of colored water, fish kills, the beaching of marine mammals and other unusual activity, or reports of human illness following consumption of fish or shellfish. When toxins show up in laboratory analyses of samples of edible species, warnings are issued and shellfish harvesting and some kinds of fishing may be halted. Economic losses can be high when commercial fishing and aquaculture operations (including fish and shellfish farms) are affected.

To better manage the human risk associated with HABs, scientists are continuing to research methods of rapid analysis to identify toxic phytoplankton species and to detect marine biotoxins in water, phytoplankton, and animals. Better monitoring can help decrease the incidence of overly conservative fishery closures by delineating the extent of the threat, thus reducing the need for broad-scale closures due to lack of information.

The Harmful Algal Bloom and Hypoxia Research and Control Act was enacted in 1998. The act recognizes that HABs threaten coastal ecosystems and endanger human health.

A national assessment, published in early 2001, recognized the threat to human health and coastal economies, but found that management options are limited. HAB impacts can be minimized through monitoring programs that regularly sample shellfish to detect HAB toxins, and issue warnings when toxins are found. Satellite remote sensing can track offshore blooms, alerting coastal communities to potential problems as blooms come inshore.

ALGAL BLOOMS IN THE OCEAN

The ocean, that vast body of water covering 71 per cent of the Earth's surface, is divided into four major basins: the Pacific, Atlantic, Indian, and Arctic Oceans. These large basins are interconnected with various shallow seas, such as the Mediterranean Sea, the Gulf of Mexico, and the South China Sea. Oceans and seas abound with life, ranging from microscopic unicellular (one-celled) organisms to multicellular (many-celled) animals.

Algae is an important life form in the ocean. Life in the ocean is maintained in balance by forces of nature and by predator–prey relationships, unless some external pressures upset the balance. When a balance upset leads to conditions more favorable for the reproduction and growth of algae, an explosive increase in the number of algal cell density occurs. Such rapid increases in the algae population are called algal blooms.

During a bloom, a liter of water may contain millions of algae. The most widely publicized type of algal bloom is associated with species that produce a toxin (chemical substance) harmful to animals that feed on the algae (and hence is known as a harmful algal bloom), and/or algae that cause a tint in the water because of the photosynthetic pigments they contain. The latter commonly is known as a "red tide", but different pigments can turn the water red, brown, purple, orange, or yellow. Depending on the circumstances and the species present, a red tide may or may not be harmful. Although not all algal blooms in the ocean produce highly visible effects nor are all blooms harmful, they nonetheless affect life in the ocean and on land in both beneficial and harmful ways.

In February 2002, the massive die-off and decay of algae from a nearshore harmful algal bloom (a "red tide") caused a rapid reduction in the water's dissolved oxygen concentration, driving tens of thousands of rock lobsters to "walk out of the sea" near the coastal town of Elands Bay in South Africa's Western Cape province. The lobsters in search of oxygen moved toward the breaking surf, but were stranded when the tide went out. Government and military staff attempted to save some of the lobsters, but others were collected for food. A similar stranding from a massive red-tide event occurred at Elands Bay in 1997.

Requirements for a Bloom

Algae require warmth, sunlight, and nutrients to grow and reproduce, so they live in the upper 60 to 90 metres (200 to 300 feet) of ocean water. The upper layer of water, the epipelagic zone, is rich in oxygen, penetrated by sunlight, and warmer than water at lower levels. As algae and other organisms that live in the ocean die, they fall to the bottom of the ocean, where they decay and release the compounds from which they were made. Under certain conditions, these nutrients can deplete the oxygen in the water.

Temperature and salt concentration determine the density of water and how water moves (currents). Cold water is denser (heavier) and sinks from the surface (downwelling). Other water moves across to replace it. Eventually, water at the surface is replaced by water that has risen, or upwelled, from the bottom to the surface somewhere else in the ocean. These upwellings bring nutrient-rich waters to the top. This increase in nutrients can trigger algae blooms.

An increase in nutrients also may be caused by activities of humans, such as runoff from animal farms or fertilized croplands and lawns, or atmospheric deposition of sulfur and nitrogen compounds or oxides derived from the burning of fossil fuel. These nutrients lead to blooms in coastal waters to a greater extent than in the open ocean.

However, some of these nutrients do find their way to the open ocean far from shore, and contribute to the formation of blooms in the open ocean. Their movement is aided by the wind and by ocean currents. Algae blooms in the open ocean are not usually harmful; instead, they provide many benefits, largely deriving from the fact that the open ocean is relatively unproductive (low in nutrients).

ALGAE AND PHOTOSYNTHESIS

Algae are referred to as plants because, like plants, they produce organic compounds from inorganic compounds (carbon dioxide and water) by capturing and using the energy from sunlight. Most algae are eukaryotic, an exception being the blue-green algae (cyanobacteria).

Photosynthesis takes place in organelles called chloroplasts in eukaryotic cells. Chloroplasts contain an outer and an inner membrane and pancakeshaped structures called thylakoids. Energy is captured from sunlight by pigments (chlorophylls *a* and *b* and carotenoids) stored in the thylakoids.

Photosynthesis occurs in two stages commonly referred to as the light reactions (light is required) and the dark reactions (no light is directly required). During the light reactions, energy captured from sunlight is used to split (dissociate) water molecules. Electrons released from this reaction are passed down a series of electron carrier molecules, leading to the storage of the energy in the form of ATP (adenosine triphosphate). This is the form in which living organisms store energy to be used immediately for carrying out chemical reactions and other activities.

Oxygen is produced as a byproduct of the light reactions. During the dark reactions (Calvin cycle), six molecules of carbon dioxide are used to make sugar (glucose). Because algae use carbon dioxide and release oxygen as a product of the light reactions, these plants play an important role in maintaining the proper concentrations of carbon dioxide and oxygen in the environment, via the carbon cycle and oxygen cycle.

Algae, like green plants, produce the first organic compounds in the food chain and thus are referred to as primary producers. Other organisms cannot use inorganic molecules to make the organic compounds that they need for life, and therefore depend on algae and other plants as the initial source of organic compounds. These organisms either eat algae to obtain organic compounds, or obtain them from the water when they are released after the algae die.

Cyanobacteria

Cyanobacteria, also known as blue-green algae, are one of the oldest known types of algae and are believed to have played a major role in the addition of oxygen to the Earth's early atmosphere. Some cyanobacteria carry out nitrogen fixation, which is the conversion of nitrogen gas into nitrogen compounds that can be used by other primary producers.

Diatoms

Diatoms are unicellular and have a cell wall composed of silica, a glass-like material, which comprises a shell-like structure called a frustule.

Coccolithophore blooms are identifiable via space-based remote sensing because their external plates of calcium carbonate, called coccoliths, backscatter light from the water column to create a bright optical effect. This bloom (the cloudy swirl in lower lefthand corner) occurred in summer 2001 in the Celtic Sea off England's southwestern coast. When diatoms die, the frustules settle to the bottom of the ocean floor and combine with the soil to form diatomaceous earth. Diatomaceous earth is used in products such as filters for swimming pools, as temperature and sound insulators, and as an abrasive in toothpaste.

Dinoflagellates

Dinoflagellates have two unequal flagella that help them direct their movement. Many of these organisms contain colored pigments that cause the water to appear colored when these organisms bloom, leading to the terms "red tide" or "brown tide," for example. Some dinoflagellates live in close association with marine animals, such as sponges, sea anemones, giant clams, and corals. The golden-brown photosynthetic cells found in these animals, called zooxanthellae, actually are dinoflagellates.

Coccolithophores

Coccolithophores are cells covered with button-like structures called coccoliths made of calcium carbonate. The coccoliths give the ocean a milky white or turquoise appearance during intense blooms. The long-term flux of coccoliths to the ocean floor is the main process responsible for the formation of chalk and limestone.

Coccolithophores and some other algae participate in the sulfur cycle and produce the gas dimethyl sulfide. This is the primary way that sulfur is carried between ocean and land.

Dimethyl sulfide leaves the surface of the water and reacts with oxygen in the atmosphere to form tiny sulfuric acid droplets. These droplets are carried over land and fall back to land in the form of precipitation. They also aid in the formation of clouds, which partially block the transmission of harmful ultraviolet light that penetrates the surface water. Cloud formation also is thought to encourage surface winds that promote the movement of surface water, leading to upwellings that bring nutrients to the surface.

Benefits of Algal Blooms

Algal blooms provide large concentrations of algae that produce organic compounds needed by higher organisms, ranging from oysters, clams, and mussels to human beings. For this reason, productivity increases in areas where algal blooms occur. More algae in the water means that more carbon dioxide is used from the atmosphere and that more oxygen is released into the atmosphere. Oxygen is necessary for many living things, including humans. As noted previously, the production of dimethyl sulfide gas helps protect algae from harmful ultraviolet rays so they remain healthy and thus are able to continue the cycle of sustaining life on Earth.

Even in the coldest parts of the ocean, algae provide the primary source of organic material to animals at the bottom of the food chain. Organic materials are moved up the food chain as higher organisms feed on those lower down the chain. For example, algae have been found in Antarctic sea ice. As sea water freezes, algae living in the water are frozen in the ice, where they later can be released during a thaw. These algae are a vital source of food for krill, the shrimp-like organisms eaten by penguins, seals, seabirds, and whales.

Ecology, Marine

Marine ecology describes the interactions of marine species with their biotic (living) and abiotic (nonliving) environments. The biotic environment includes interactions with other living organisms. The abiotic environment includes aspects of the physical habitat, such as water temperature, chemical composition, depth, and current.

Trophic Levels and Biomass Pyramids

The word "trophic" refers to feeding, and "trophic levels" describe the feeding levels in a food chain. The first trophic level includes species known as primary producers. These organisms produce organic material from

Krill, small crustaceans that frequently are found in dense swarms, are critical to the food chain in polar oceans. Marine mammals and penguins, for example, rely on krill for a major part of their diet. inorganic substances using resources from the environment and an external source of energy.

Most primary producers are photosynthesizers—that is, they use sunlight as their energy source. Plants and algae are examples of photosynthesizers. Nearly all food chains are based on photosynthesizers, although a few marine food chains depend on bacterial chemosynthesizers, which use a chemical source of energy for production. Chemosynthesizers are discussed later in this entry.

The next level in the food chain, the second trophic level, consists of species that eat the producers. These are sometimes referred to as primary consumers or as herbivores (plant-eaters).

The third trophic level consists of secondary consumers, which are also called carnivores (animal-eaters). There can be further, higher trophic levels as well.

Finally, there are detritivores and decomposers, both of which feed on dead or decaying organic matter. Much of the decomposition work in food chains is done by bacteria.

Species can occupy more than one trophic level, and each trophic level usually has many representatives. Consequently, in most marine ecosystems, trophic interactions are described not by simple chains but as complicated food webs. Many species, including omnivores (eaters of both plants and animals), eat at more than one level of the food web.

Biomass Pyramid

A pyramid of biomass describes the total amount of biomass, or weight of living matter, that is present in each trophic level. The amount of biomass decreases sharply as one moves up from one trophic level to the next (thus the "pyramid"). That is, the total weight of all the producers in a food chain is greater than the total weight of all the primary consumers, which in turn is greater than that of all secondary consumers. This is because not all the energy that a consumer obtains from food is converted to new biomass.

For example, the consumer must use energy to catch and eat prey. In addition, a lot of energy is lost to metabolism and heat. In fact, only about 10 percent of the energy in one trophic level is passed on to the next level. Because of the decreasing amount of energy available in each trophic level, most ecosystems cannot support more than four or five trophic levels.

Phytoplankton and Zooplankton

In nearly all marine ecosystems, photosynthetic species represent the producers at the base of the food chain. Photosynthesis requires sunlight, carbon dioxide, and nutrients such as nitrogen and phosphorus, which are found in sea water.

Although there are some photosynthetic rooted plants in shallow marine areas, the majority of photosynthetic organisms in the ocean are microscopic algae, or phytoplankton, that drift along with currents in the water. Phytoplankton are found only in the topmost layer of marine water, known as the epipelagic zone, where there is enough sunlight for photosynthesis. Because of the concentration of producers, many consumer species also are found close to the w. surface.

The amount of phytoplankton in oceans varies across regions and also changes seasonally. For example, phytoplankton are found in low concentrations in tropical waters, where nutrients are in short supply. Phytoplankton density generally is lowest during the winter, when resources are scarce, and greatest during the spring, when levels of sunlight and nutrients increase.

Spring often brings algal blooms, or population explosions of phytoplankton. Algae can be so plentiful during the blooms that they color the water, particularly if they contain red, brown, orange, or purple pigments. The amount of phytoplankton available has implications all the way up the food chain, and blooms in algae populations are often followed by increases in populations of other species.

Phytoplankton are consumed by many organisms, including diverse species of zooplankton. Zooplankton are free-floating consumers and include single-celled protozoa, tiny crustaceans, and the larval stages of species such as mollusks and fish. Zooplankton and phytoplankton are consumed by the nekton, free-swimming marine organisms such as fish, marine mammals, and penguins, as well as species on the ocean bottom, including bivalves, crustaceans, and snails.

Examples of Marine Ecosystems

There are numerous types of marine ecosystems. These include coral reefs, tidepools, polar oceans, the abyss, and others.

Coral Reefs

Photosynthetic algae are the producers in coral reefcommunities. The coral reefs represent one of the most diverse marine communities—in fact, a quarter of all marine species are found in or near coral reefs.

Photosynthetic algae in coral reefs have a mutualistic relationship with coral, that is, a close association that benefits both members. The algae are sheltered within the corals' calcium carbonate shells and provide nutrients to the corals in exchange. Diverse invertebrates feed on algae and in turn are eaten by reef fish, which are eaten by larger fish species.

Tidal Pools

Tidepool ecosystems are ones that are alternately submerged by water (at high tide) and exposed to air (at low tide). Most of the species found in tidepools are unique to the habitat and are able to survive in both wet and dry conditions. Tidepool food webs, like most marine food webs, are based on algae. Consumer species include bivalves, snails, small fish, and sea anemones. Often the top predators in tidepools are starfish.

Polar Oceans

In polar habitats, such as the oceans around Antarctica and the Arctic, photosynthetic algae also form the basis of the food chain. Algae are eaten by shrimp-like crustaceans called krill, which in turn serve as food for species as diverse as penguins and baleen whales. Penguins are eaten by seals, and penguins and seals are eaten by some whales, particularly killer whales, which are the topmost predators in the system.

Kelp Forests

Kelp forests are marine ecosystems characterized by gigantic species of floating photosynthetic algae called kelp. Kelp can reach lengths of up to 80 meters (about 262 feet). A number of crustaceans feed on kelp, as do sea urchins. These are eaten by larger organisms, such as sea otters.

The Abyss

The deepest part of the ocean, the abyss, extends to depths as great as 6,000 meters (about 19,685 feet, or 3.7 miles). These deep-sea environments are characterized by cold temperatures and lack of light. Consequently, no photosynthesizers exist, although a diverse array of detritivores feed on dead organic matter that floats down from above. There are also numerous predatory and parasitic species.

Hydrothermal Vents

Hydrothermal vents are cracks or openings in the ocean floor where hydrogen sulfide, metals in solution, and other chemical compounds escape into the sea water.* Certain specialized chemosynthetic bacteria live in these hot areas and produce organic matter from hydrogen sulfide. They form the base of unusual food webs in this specialized habitat.

Chemosynthetic bacteria are eaten by specialized vent worms, clams, and mussels, which in turn provide food for octopuses and other species. The hydrothermal vent crab is at the top of the food chain in vent environments.* These unique deep-sea ecosystems are more diverse than most deep-sea environments.

Bioconcentration

A feature of the ordered structure of food chains is that substances, especially pollutants, become concentrated in large amounts at higher trophic levels. This process is called bioconcentration.

Bioconcentration occurs because species that consume pollutants do not excrete them but, rather, store them in bodily tissues, where they accumulate over time. For organisms high in the food chain, their prey organisms have already concentrated pollutants from multiple prey of their own, and so forth. This results in high concentrations of pollutants in species that eat high in the food chain.

An issue of particular relevance in aquatic ecosystems is mercury poisoning. Mercury causes severe health problems in human and other species, including brain damage, and is bioconcentrated in the upper trophic levels. This is why people, pregnant women in particular, are often advised against eating predatory fish such as tuna and swordfish.

CORALS AND CORAL REEFS

A coral reef is a structure in the sea constructed by coral skeletons and limestone debris that remains in place after the plant or animal dies. The structure is geological, the communities include plants and animals, and they are controlled by meteorological and oceanographic conditions.

Distribution and Roles of Coral Reefs

Coral reefs include shallow-water tropical reefs, such as the Great Barrier Reef off the east coast of Australia, and deep-water reefs, such as Sula Bank off Norway. Tropical reef diversity (the number of different plants and animals) is high, with complexity similar to tropical rain forests.

Natural and Human Values

Coral reefs function as living breakwaters. For example, on Bikini Atoll, Marshall Islands, the wave energy that strikes this reef amounts to 20 horsepower per surge channel per wave (waves arrive at the surge channel at a rate of one wave approximately every 8 seconds).

A cross section of a reef shows that it is built by multiple generations of corals and that there is significant void space. This empty space provides refuge for plants and animals. A coral cut in half would reveal small tunnels and galleries created by boring sponges, clams, and other animals that provide homes for worms, shrimp, and fish.

The tourism and fisheries industries supported by coral reefs are important to the economies of tropical areas, as well as the economies of the countries from which the tourists travel. Conservation efforts have been aided by the high visibility of reefs and their biocommunities.

Coral: Simple Animals

Corals exist at the tissue level: they do not have organs, such as a heart. On the evolutionary ladder, corals are one step above the sponges. They are the simplest animals to have nervous and connected muscular systems and a dedicated reproductive system. They belong to the phylum Cnidaria, which also includes the jellyfish.

Tetiaroa in French Polynesia's Society Islands is an atoll reef that formed at the edge of an old submerged volcano. It forms an irregular ring around a shallow central lagoon, and its outside edges drop steeply to the ocean floor.

Tentacles and sticky mucus help coral polyps trap plankton. Coral are simple animals in the same phylum as jellyfish.

Each coral animal consists of an individual sac-like body called a polyp. Tentacles radiate from the mouth end of the polyp and bristle with tiny hairs; these are the triggering devices for a microscopic harpoon called a nematocyst. The harpoon includes a barb for injecting neurotoxin, and a tube to connect the barb and the toxin's reservoir. Corals are colonial animals, and multiples of polyps form a colony.

Types of Coral

The more common corals include hydrocorals, octocorals (polyps with eight tentacles), and scleractinian corals (polyps with six, or multiples of six, tentacles).

Hydrocorals

The hydrocorals include the fire corals (genus *Millepora*). *Millepora* is common on shallow, tropical–subtropical reefs in the Caribbean and the Pacific. Hydrocorals have a life cycle of alternating sexual and asexual generations known as metagenesis. The hydroid asexual generation attaches itself to the reef and produces a medusa offspring by budding. The medusa generation sends eggs and sperm into the water, and the product of this union, a free-swimming larva, settles on the bottom and grows into a new hydroid coral. The upward growth for *Millepora* is about one centimetre (2.5 inches) annually.

Octocorals

Octocorals include sea fans, plumes, mats, and rods. They are flexible and sway to and fro in the waves. The octocoral skeleton is a matrix of limestone structures called spicules and an organic connective material. Spicules provide strength like bones, and the connective material holds the spicules together.

Octocorals reproduce sexually, with eggs and sperm released in the water column. Following fertilization, the larvae may remain in the water column for days or weeks before settling and attaching to the sea floor. An individual larva settles and grows into an adult. Octocorals in the western Atlantic grow from 1 to 4 centimetres (2.5 to 10 inches) annually. Scleractinian Corals. The scleractinian corals have many growth forms: branching, hemispheres, columns, sheets, mushrooms, and tubes. Because of the strength and persistence of their limestone skeletons, these corals are the principal reef architects.

Reproduction in the scleractinian corals is either by broadcasting eggs and sperm into the water or by internal fertilization and brooding of larva. Some corals have female and male sex organs in the same polyp and are capable of self-fertilization. Larvae either live in the water column or crawl along the bottom.

Branching corals have relatively rapid growth (15 centimeters, or 6 inches, per year). The boulder corals grow 1.2 to 2.5 centimetres (0.5 to 1 inch) per year. Branching corals are more fragile, often breaking during storms and generating fragments that in turn can grow into new coral colonies. Boulder corals generally do not fragment.

Symbiosis: Living Together

Nearly all the shallow-water corals and related Cnidarians have a microscopic symbiotic algae living in the tissues. The alga, called zooxanthellae, is an important partner in the success of tropical coral reefs. Zooxanthellae provide the means for corals to sustain high growth and reproduction rates in waters that are low in nutrients. Metabolic wastes generated by animal tissues are used by the alga; the corals use the fats, oils, and sugars synthesized by the alga during photosynthesis.

For the process to function optimally, three things are necessary: sufficient light, clear water, and temperatures that range from 20 to 30°C (68 to 86°F). The process of photosynthesis is conceptualized as follows:

Carbon Dioxide + Water light energy ? Sugar + Water + Oxygen + Energy

The energy provided by photosynthesis enhances coral growth. Corals grow by taking in calcium ions from the sea and combining them with bicarbonate ions. The result, calcium carbonate, bonds as a single crystal onto other crystals, creating the limestone skeleton. Energy generated from photosynthesis expedites the calcium carbonate movement from the tissues to the skeleton. The steps follow:

Calcium bicarbonate formation: $Ca^2+ + 2HCO_3- ?$ $Ca(HCO_3)_2$

Calcium carbonate and carbonic acid formation: $Ca(HCO_3)_2 ? CaCO_3 + H_2CO_3$

Carbonic acid ionization: $H_2CO_3 ? H^+ + HCO_3^-$

Conversion to water and carbon dioxide: $H_2CO_3 ? H_2O + CO_2$

Examples of Coral Reef Types

Reefs in the Pacific Island chains (Hawaii, French Polynesia, and the Marshall Islands) initially are formed on the sides of old volcanic mountains and are called fringing reefs. In the process of tectonic plate movement, the plate carries the mountain into deeper water, and the reef becomes a ring that surrounds the mountain but is separated from it by a body of water called a lagoon. The reef is now called a barrier reef. Over thousands of years, the mountain submerges, leaving a ring of reef and a few low islands called an atoll. The terms "fringing", "barrier", and "atoll" were first used in 1834 by Charles Darwin in his book about coral reefs.

Norway

Deep-water coral assemblages include those off the coast of Norway. Because they live in darkness, deep-water corals do not have the Coralline algae produce calcium carbonate skeletons that assist in building reefs. The coralline algae typically are found in areas with heavy wave surge. symbiotic zooxanthellae that are common to the shallow reefs. The reefs are built up from the sea floor by a branching coral, *Lophelia pertusa*.

These coral banks are often associated with petroleum gas seeps that are surrounded by bacterial mats. The biological productivity (chemosynthesis) from these mats supports a food web somewhat similar to deep-sea thermal vent communities: high biomass and productivity without the influence of light. Corals, presumably, obtain small prey animals (crustaceans, worms, and mollusks) that feed on the bacterial mats.

Lophelia pertusa banks are often miles long, up to 30 meters (100 feet) high, and equally wide. The maze of twisted and interlocking branches provides refuge for resident fish populations.

Nova Scotia

Off the coast of Nova Scotia, on the fishing banks, most corals are large octocorals; some old and large colonies are up to 5 meters (about 16 feet) high. *Paragorgia arborea,* a common coral, is pink to orange in color with branches ending in blunt bundles resembling a wad of gum, living up to its common name—bubblegum coral. Although this coral is not well studied, its presumed ecological value is as a refuge for juvenile fish. This is a region that in the past had large populations of codfish; however, overfishing has decimated the fishery.

Florida

Off the central east coast of Florida, another branching coral, *Oculina varicosa,* builds banks upward from the sea floor. *Oculina* banks are a refuge area for a variety of fish, such as the snowy grouper. These deep-water coral reefs exemplify two common themes of deep-water corals: one species builds the framework, and photosynthesis is not a major energy source.

Physical and Biological Controls

Tropical cyclones are the most important natural force controlling reef development. Storms generate massive waves that break up coral formations, heavy rains that reduce salinity, and silt deposits in reefs close to high.

The crown-of-thorns starfish (or acanthaster) eats coral polyps. If present in sufficient numbers, this voracious starfish can decimate coral populations. islands. Damage and recovery depends on storm strength, wind direction, duration, and the frequency of events.

Crown-of-Thorns Starfish.

The biological controls on coral reefs generally are not as catastrophic as a tropical cyclone; however, a number of reefs have been virtually picked clean of coral thanks to a voracious predator. In the Pacific and Indian Oceans, a starfish called the crown-of-thorns *(Acanthaster planci),* has a history of population explosions. When the crown-of-thorns (COT) reaches abundances of several per square meter, they will eat virtually every coral in sight.

In the 1970s, conventional wisdom held that COT outbreaks were related to insufficient predators to control them. In Australia and Guam, a bounty of $1 per COT collected was instituted. Others placed the blame on pollution.

However, geologists sampled deep sediments on the reefs and found pockets of COT spines. Determining the age of these spines indicated that COT population explosions occurred thousands of years before the arrival of Europeans in Australia. The COT outbreaks seem to be related to the typhoons that occasionally strike the east coast of Australia. The torrential rainfall from the typhoon causes heavy flooding, bringing nutrients into the near-shore waters around the Great Barrier Reef.

The COT starfish usually lays millions of eggs, which develop into floating (planktonic) larvae. Normally, the larvae would have a poor chance for survival because of a lack of food. After a flood event with the enriched water, phytoplankton blooms occur. Such blooms are key to COT larval survival. The COT larvae thrive on the rich soup of plankton, and when they metamorphose into the adult COT, huge swarms of COT begin feeding on coral.

If a COT event occurs approximately once per decade, there is sufficient time for the coral to recover before the next

COT disturbance. However, if a COT event occurs every three years, the reef does not have sufficient time for the coral larvae to settle and grow and replace the lost corals.

Long-Black-Spine Sea Urchin

The concept of keystone species for coral reefs is not subscribed to universally; however, there is some validity to the idea. In western Atlantic coral reefs prior to 1983, a prodigious algae grazer was held up as the keystone for controlling the algal growth on the reefs and for creating space for the coral larvae to settle. Studies have documented the value of the long-black-spine sea urchin in removing algae and enhancing coral recruitment on Caribbean reefs.

In 1983 and 1984, a pandemic disease killed 99 percent of the urchins on reefs from Barbados to Panama. Since that disease event, the urchin has not recovered to reclaim its role as a major herbivore. Today, most Caribbean reefs have moderate to high algal cover. There is continuing debate as to whether the algal cover seen today is the result of higher levels of nutrients in the waters, or whether it is due to the lack of grazing.

Hazards and Global Stresses

Various environmental hazards can threaten coral reefs. Some stresses are natural, whereas others are magnified by human activities.

Thermal Stress

Elevated surface sea-water temperatures cause coral bleaching at the time of seasonal maximum heating (late summer to early fall). This thermal stress results in the expulsion of the zooxanthellae (symbiotic alga), causing the corals and similar reef inhabitants to turn white and, in severe cases, to die. Loss of the zooxanthellae results in reduced or no growth, no reproduction, and susceptibility to disease.

Corals in shallow tropical reefs are at the near-margin of upper thermal tolerance. Most corals will tolerate water of 30°C (86°F); however, if the temperature reaches 32°C (90°F), corals will bleach.

Factors that affect temperature include tide, wind, time of year, cloud cover, and currents. In the 1960s and 1970s, bleaching events were rare, occurring about once a decade in a region. By the 1990s, bleaching events were occurring every other year or every third year. The magnitude of the events (oceanwide and the extended duration of the episode) resulted in massive coral mortalities in reefs in the Indian Ocean and western Pacific. An El Niño event could have played a major role in a vast coral bleaching event in French Polynesia's Rangiroa Atoll in 1998.

Global temperature is rising, and this in turn elevates surface sea-water temperatures. Global warming is believed to be the result of increased greenhouse gases (such as carbon dioxide) and an increase in ultraviolet radiation due to the loss of ozone in the atmosphere caused by the release of chlorofluorocarbon compounds (CFCs).

Dust

Satellite images show the movement of dust across the oceans from other continents.* African dusts from the sub-Saharan region are entrained in the upper winds and carried across the Atlantic. The dust includes spores from bacteria and fungi, and mineral elements such as iron, copper, mercury, and arsenic. A well-documented disease in a Caribbean sea fan *(Gorgonia ventalina)* is caused by a fungus whose spores are carried in African dust. The ocean, land, and atmosphere are interconnected; no part of the global community is insulated from influences that come from other continents or oceans.

Fish and Agriculture

Reefs are under siege in some places by destructive fishing practices: dynamite, chemicals, dredging, poor land-use practices, slash-and-burn agriculture on hillslopes, and fishing efforts beyond maximum.

Tropical settings and a diversity of easily viewable biota make snorkeling a popular tourist activity on shallow-water coral reefs. Deep-water reefs are less accessible and hence have not captured as much popular appeal. sustainable yield.

The recent trend in protecting reefs is to designate large areas as parks or reserves and not to allow the harvest of any plant or animal. This maximizes conservation of the entire reef community.

Preserving Coral Reefs

Marine protected areas (MPAs) permit visitors to look at, but not take, anything; these areas therefore are a reasonable way to protect reef resources. An MPA protects the habitat and the target species that the fishers are harvesting.

The Florida Keys National Marine Sanctuary and the Great Barrier Reef Marine Park in Australia have created "no-take" zones to protect all the plants and animals. The spillover effect provides a source of animals to replenish other areas; that is, as the MPA becomes saturated, some animals move away. Within two years of creating the Florida Keys no-take zones, populations of grouper, snapper (fish), and lobster increased dramatically.

LIFE IN WATER

Life is thought to have originated in an aquatic environment—the oceans. Living organisms have since adapted to numerous aquatic habitats, both.

Sea nettles and jellyfish are pelagic and planktonic, meaning that they live in the water column and primarily float or drift as opposed to swimming. marine and fresh-water. They occupy environments as diverse as lakes, rivers, and oceans.

Marine Environment

About 17 per cent of known biological species live in oceans. Marine species are described as either pelagic or benthic. Pelagic organisms live in the water column itself. Benthic species live on the ocean bottom.

Pelagic

Pelagic organisms include plankton, which float along with currents, and nekton, which are active swimmers.

Plankton are divided into phytoplankton, which include photosynthesizing species such as algae, and zooplankton, which are consumer species. Zooplankton consist largely of copepods (tiny crustaceans).

Although plankton generally drift with ocean currents, some plankton have limited mobility. For example, certain zooplankton species move towards the water surface at night to feed, when there is less danger of predation, and return to deeper waters during the day.

Organisms that are planktonic throughout their life cycles are known as holoplankton. Organisms that are only planktonic during the early parts of their life cycles are called meroplankton. Meroplankton include the larval or juvenile forms of many species of fish and mollusks. These species use the planktonic stage to disperse to new areas. Although most planktonic species are small, some are large, such as kelp and jellyfish.

Nekton are active swimmers that use diverse means to propel themselves through the water. Some species swim using fins, tails, or flippers. Other species, such as mussels, move by shooting out jets of water, known as jet propulsion. Nektonic species include fish, octopus, sea turtles, whales, seals, penguins, and many others. Many nektonic species eat high in the food chain, although there are plankton-eating species (e.g., some fish) and herbivorous species (e.g., sea turtles) in addition to carnivorous ones (e.g., seals and killer whales).

Pelagic marine species may also be categorized according to the depths at which they occur. Different water depths are characterized by differences in temperature, amount of sunlight received, and availability of nutrients. The epipelagic zone describes oceanic waters closest to the surface, and is the zone richest in marine life. In the epipelagic zone, there is enough sunlight for photosynthesis. For that reason, the epipelagic zone is also called the photic (light) zone. All photosynthetic species, including the phytoplankton, live in this zone, as do many of the species that feed on phytoplankton.

Below the photic zone is the aphotic zone, which is characterized by very limited light (or no light) and limited food. Species in the aphotic zone often depend on food drifting down from above. Consequently, there are many detritivores (species that feed on dead or decaying organic matter) in these habitats.

Certain deep-sea habitats can be highly diverse. In the deep-sea vents, for example, chemosynthetic bacteria (rather than photosynthetic species) form the basis of the food chain. These bacteria obtain energy from chemical sources such as hydrogen sulfide instead of from sunlight.

The seahorse is pelagic and nektonic, meaning that it lives in the water column and swims rather than floating or drifting.

Benthic

Benthic species live on the ocean bottoms, and represent the greatest proportion of marine species; in fact, 98 per cent of marine species are benthic. Benthic species include epifauna, which live on the surface, and infauna, which burrow into seafloor sediment. Benthic epifauna include species such as oysters, scallops, sea stars, crabs, and lobsters. Examples of infauna include clams and many species of worms.

Some benthic species are sessile (non-moving), and live attached to the ocean bottom. Benthic plants generally are found only in shallow waters where there is enough sunlight for photosynthesis. However, benthic animals are found at a wide variety of depths, including in the deepest parts of the ocean. Some species, such as flounder, are capable of both benthic and nektonic existence.

Distance from Shore

The distance of a zone from shore can categorize marine environments. The neritic zone describes coastal marine regions. The neritic zone is particularly rich with life because the relatively shallow water allows for plentiful photosynthesis, and because a steady flow of nutrients is washed into the water from land.

Farther from land, areas of open ocean are described as the oceanic zone. The oceanic zone has significantly less total **biomass** than the neritic zone. The intertidal zone is the area of shore that alternates between being submerged and dry, depending on the level of the tide. Numerous species are specialized for living in the intertidal zone, including the familiar barnacles. The intertidal zone has the greatest density of living organisms among marine environments.

FRESH-WATER ENVIRONMENT

Fresh-water habitats are extremely diverse, and include both still-water environments like lakes and ponds, and flowing-water environments like rivers and streams.

Still-Water Habitats

Like oceans, lakes have pelagic and benthic zones. The temperature of lake water varies depending on depth, and can also change dramatically over seasons. The epilimnion is the topmost layer of lake water. It is significantly warmer than deeper areas due to heating by sunlight. The hypolimnion layer describes deeper, colder lake water. Many of the nutrients in lakes collect at lake bottoms.

Turnover occurs when all the water in a lake is nearly thermally uniform and mixed, distributing nutrients throughout the water. Turnover occurs twice a year in many temperate lakes, but may occur only once in subtropical environments, or not at all in permanently stratified lakes.

Lakes also can be described as either oligotrophic or eutrophic (or in between these two extremes). Oligotrophic lakes have low levels of nutrients and low productivity. They generally contain cold, highly oxygenated water and support species adapted to these conditions. Eutrophic lakes, on the other hand, have plentiful nutrients and are highly productive. Species that inhabit eutrophic lakes must be tolerant of low oxygen levels and warm temperatures. In general, oxygen levels in lakes depend on the amount of water circulation, the surface area that is exposed to air, and levels of oxygen consumption by living organisms.

The sea anemone anchors to the bottom substrates: in this case, the volcanic rocks of Hawaii. Anemones are classified as a benthic epifauna: that is, living on the surface of the ocean bottom rather than being burrowed.

Quatic snakes include amphibious species that live both on land and in the water and species that occupy marine habitats only (such as this yellow-bellied sea snake). Some have paddle-like tails that help propel them through the water. All aquatic snakes breathe using lungs, however, and must return to the water surface to obtain oxygen.

Flowing-Water Habitats

River habitats are characterized by flowing water. River species generally have special adaptations for living in water currents. Some species are sessile and live anchored to the river bottom. Other species have evolved adaptations such as suckers or hooks to keep themselves from being washed away. Still other species are strong swimmers. Many of these have flattened bodies that help them resist the pressure of the current.

Compared to lakes, rivers tend to be well-oxygenated because of the constant motion of the water. Temperatures can change quickly in rivers, but do not span as great a range as in lakes or other still water. Because there is less penetration of light in flowing water, plant diversity is generally lower in rivers than in lakes. As in other aquatic ecosystems, algae frequently occupy the base of the food chain.

Challenges of Aquatic Life

Flotation

Flotation or placement in the water column is a challenge faced by all aquatic organisms. For example, it is crucial to phytoplankton to stay in the photic zone, where there is access to sunlight. The small size of most phytoplankton, plus a special oily substance in the cytoplasm of cells, helps keep these organisms afloat.

Zooplankton use a variety of techniques to stay close to the water surface. These include the secretion of oily or waxy substances, possession of air-filled sacs similar to the swim bladders of fish, and special appendages that assist in floating. Some zooplankton even tread water.

Fish have special swim bladders, which they fill with gas to lower their body density. By keeping their body at the same density as water, a state called neutral buoyancy, fish are able to move freely up and down.

Salinity

Aquatic species also have to deal with salinity, the level of salt in the water. Some marine species, including sharks and most marine invertebrates, simply maintain the same salinity level in their tissues as is in the surrounding water. Some marine vertebrates, however, have lower salinity in their tissues than is in sea water. These species have a tendency to lose water to the environment. They make up for this by drinking sea water and excreting excess salt through their gills.

Fresh-water aquatic species have the opposite problem—a tendency to absorb too much water. These species must constantly expel water, which they do by excreting a dilute urine.

Species that occupy both fresh water and marine habitats at different stages of their life cycle must transition between two modes of maintaining water balance. Salmon hatch in fresh water, mature in the ocean, and return to fresh-water habitats to spawn. Eels, on the other hand, hatch in salt water, migrate to fresh-water environments where they mature, and return to the ocean to spawn.

Salinity is particularly variable in coastal waters, because oceans receive variable amounts of fresh water from rivers and other sources. Species in coastal habitats must be tolerant of salinity changes and are described as euryhaline. In the open ocean, salinity levels are generally constant, and species that live there cannot tolerate salinity changes. These organisms are described as stenohaline.

Temperature

Eurythermal species are those that can survive in a variety of temperatures. Eurythermality generally characterizes species that live near the water surface, where temperatures change depending on the seasons or the time of day. Species that occupy deeper waters generally experience more constant temperatures, are intolerant of temperature changes, and are described as stenothermal.

PLANKTON

Awareness is growing regarding the importance of the oceans and the variety of life they support. Research in many branches of oceanography is discovering the vast unknown of the marine world, and has expanded interest in the understanding of the marine environment and the role each member plays in a complex community.

The free-floating organisms known as plankton, from the Greek "wandering," are the drifters of the ocean. Although most of these organisms are motile (moving), they cannot swim or move against currents, but they can move vertically in the water column.

Many marine plankton are found in the deep waters of the outer ocean, or pelagic waters, whereas others are found in the shallow waters known as the neritic zone. Many of the neritic plankton are known as meroplankton, and spend only a brief period of their life cycle in the planktonic category. Many pelagic forms, such as the holoplankton, are planktonic during their entire lifespan.

The size of plankton can also determine its general name.

- *Picoplankton:* Smaller than 2 µm; includes bacteria, prochlorophytes, and viruses.
- *Nanoplankton:* 2 to 20 µm; includes diatoms, coccoliths, and silicoflagellates.
- *Microplankton:* 20 to 200 µm; includes large diatoms, dinoflagellates, and small zooplankton, such as ciliates.
- *Macroplankton:* 200 to 2,000 µm; includes large zooplankton, copepods, and invertebrate larvae.

- *Megaplankton:* Larger than 2,000 μm; includes fish larvae and gelatinous zooplankton.

PHYTOPLANKTON

Many kinds of marine and fresh-water organisms utilize inorganic carbon (as carbon dioxide) and fix it into organic compounds by photosynthesis.

This vial of plankton illustrates the small sizes: some are barely visible to the naked eye, and some are visible only through a microscope. Plankton form the basis of aquatic food webs, including the ocean. The principal taxa of microscopic planktonic producers, primary producers, are found over most of the world's oceans, lakes, rivers, and estuaries, and comprise the base of the food web. Phytoplankton consist primarily of diatoms, dinoflagellates, coccolithophorids, silicoflagellates, bacteria, and viruses. All of the organisms discussed below are key players in the microbial food web.

Diatoms

Diatoms have cell walls of silica and pectin, and float in the water column or attach to surfaces as single cells or chains. They are one of the major contributors to primary production in coastal waters, and occur everywhere in the ocean, but are most abundant in colder, nutrient-rich, nearshore waters. Cell division occurs by fission, which is accompanied by a reduction in cell size. They are one of the principal groups that fix carbon through photosynthesis, and this production is prominent during seasonal blooms of short duration.

Dinoflagellates

Dinoflagellates occur as single cells, either naked or within a cellulose cell wall, and many species use flagella to move. These organisms are sometimes classified as protozoa and algae because of their ability to photosynthesize and also absorb nutrients by being parasitic, or by ingesting organic particles. They are second to diatoms in contributing to primary production, and are widespread in the oceans, but are most abundant in nutrient-poor waters offshore. Reproduction is by cell division. Some species are bioluminescent (emitting a pale blue glow seen at night). Dinoflagellates often are the cause of red and brown tides, so named because the algal pigments give the water a colored tint.

Coccolithophorids

Coccolithophorids are single-celled organisms. Many are flagellated, and are protected by ornate calcareous plates, called coccoliths, embedded in a gelatinous sheath that surrounds the cell. These organisms may form cysts that produce spores to produce new individuals. They are most abundant in warm, open-ocean waters, and are sometimes found nearshore. Coccolithophores can photosynthesize (autotrophic) and may also absorb organic matter (heterotrophic).

The microscopic phytoplankton shown here are mainly comprised of diatoms and dinoflagellates. Other phytoplankton in the microscopic ranges include coccolithophorids, silicoflagellates, bacteria, and viruses. Some phytoplankton are much larger.

Silicoflagellates

Silicoflagellates occur as single flagellated cells and typically secrete a silicious outer skeleton. Like coccoliths, these organisms are both autotrophic and heterotrophic, and are most abundant in cold, nutrient-rich waters.

Bacteria

Bacteria are prokaryotes with cell walls made of chitin, and occur as single coccoid cells or long filaments. They often are restricted to waters with low oxygen, and are important in the metabolism of aquatic ecosystems. To support their metabolism, they obtain nutrients by the uptake of organic matter and the release of exoenzymes to lyse (distintegrate or dissolve) particulate organic matter, and attack diatoms, dinoflagellates, and flagellates. Blue-green algae, or cyanobacteria, are photosynthetic. Bacterial activity in marine waters is strongly affected by availability of nutrients and organic matter. Their productivity increases as phytoplankton productivity increases.

Viruses

Viruses play an important role in marine food webs. They infect a wide range of hosts, including bacteria and phytoplankton. They can potentially reduce phytoplankton and bacterial production by viral lysing of their cells and the releasing of dissolved organic carbon. This dissolved carbon can than be utilized by other phytoplankton cells.

Prochlorophytes

Prochlorophytes are a recently discovered group of extremely abundant producers that are barely visible by microscopy. They are most abundant at the lower layers of the illuminated region of the water column, and are now considered to be another major player in primary production.

Nanoflagellates

Nanoflagellates are both autotrophic and heterotrophic. They feed on viruses, bacteria, and some picoplankton and nanoplankton. Nanoflagellates are major consumers of bacteria; some experiments show that they may be able control their abundances when larger predators, such as dinoflagellates, are not present. However, this is less likely to occur in nature.

Protozoans

Nanoplanktonic and microplanktonic protozoan groups are mainly ciliates and heterotrophic dinoflagellates. They consume bacteria, nanoplankton, and microplankton. While these groups engulf their prey, they also release nutrients that stimulate the growth of these same prey.

ZOOPLANKTON

Zooplankton are planktonic free-floating animals in fresh and marine aquatic systems, and are the major consumers of the organisms in the microbial food web. These organisms possess a wide range of feeding strategies, from the nematocysts (stinging cells) of cnidarians (e.g., jellyfish) to the complicated mouthparts of copepods. Some are carnivorous (animal-eaters), some are herbivorous (plant-eaters), and some are omnivorous (eaters of plants and animals).

These animals can move by means of cilia, flagella, jointed appendages, jet propulsion, or tailed larvae (as in tunicates to larval fish). Reproduction varies from asexual, to fission and fragmentation, to sexual reproduction where some gametes are released into the water and fertilized, yet others are retained and fertilized internally.

Zooplankton include many phylum, and not all can be discussed here. Some live their entire life cycle in the water (holoplankton), whereas only the larval stages of fish and other benthic organisms (such as starfish) live in the water column for a short time (meroplankton). All are considered zooplankton. An overview of the major zooplankton phyla follows.

Protozoa

Discussed previously, this group includes ciliates, dinoflagellates, foraminifera, and radiolarian.

Coelenterata (Cnidaria)

Typically known as jellyfish, the major groups are Hydrozoa, Scyphozoa, and Anthozoa. The hydrozoans medusae are the prominent members in zooplankton, and the most common forms are aurelia, pelagia, and siphonophores. These gelatinous animals are major consumers of smaller zooplankton and some of the microbial food web.

Ctenophora

Best known as comb jellies, these possess eight "comb" rows of fused cilia. When they are abundant, these animals can consume phytoplankton and zooplankton, and can clear the water of food for other zooplankton.

Chaetognatha

Known as the arrow worm, this is a common member of deep-water plankton. Smaller species are found in coastal waters, whereas larger species are abundant offshore in blue water. They are predacious carnivores that grasp their prey and paralyze them before ingesting them.

Annelida

This includes many species of marine polychaetes. Many of these organisms can be seen on the surface at night, shedding gametes for sexual reproduction. Their larvae are abundant in the zooplankton community.

Mollusca

This includes marine gastropod larvae, pteropods, and cephalopods (commonly known as squid and octopus).* Mollusks are consumers of larger zooplankton.

Echinodermata

This includes starfish, brittle stars, and sea cucumber. All these animals are meroplankton. Their larvae are a major presence in the zooplankton community.

Arthropoda

These are the major members of zooplankton and include copepods, shrimp, crabs, lobsters, amphipods, crustaceans, and euphausids, or krill, which are the major source of nutrition for some whales.* The most studied of crustacea are the copepods. These animals are found in all parts of the world's oceans, lakes, and estuaries and are considered the major consumers of most of the organisms in the microbial loop. Because they are holoplankton, spending their entire life in water, they can consume a wide range of food particles, from nanoplankton to microplankton, as they mature. Copepods are responsible for much of the carbon energy transferred from phytoplankton to larger zooplankton.

Chordata

Known as the urochordates (tunicates), this includes ascidians (or sea squirts) and are found on the coast, whereas larvaceans, oikopleura, thaliaceans, salps, and doliolids are pelagic and spend their entire life cycle in the water column. Tunicates are now realized to be major consumers of phytoplankton and smaller zooplankton, and can contribute to the entire food-web dynamics as much or even more than copepods.

Larvaceans have retained their notochord and tail as adults and produce a mucus net, or "house," around their bodies to capture food particles. The house is either ingested or abandoned.

The salps and doliolids are free-swimming tunicates with a cylindrical or barrel-shaped body with up to eight muscle bands to aid in swimming by jet propulsion and feeding with an internal mucus net. These animals have a complicated life cycle that includes a sexual stage and one or two asexual stages. They are known for their ability to create "blooms," or a rapid increase in their abundance, exceeding 1,000 animals in a cubic meter of water in a short period of time. With this rapid increase in population and their ability to filter feed a wide range of food sizes, they can outcompete copepods during these bloom events. Their role in the food web is being studied more intensely because of their production of large, fast-sinking fecal pellets that can transfer organic matter produced by primary producers to fish and benthic organisms.

The aquatic environment is shaped by complex interactions among a variety of physical, chemical, and biological factors. For example, physical factors such as climate, land topography, bedrock geology, and soil type influence the amount of water flowing in streams and discharging to lakes, as well as the types of materials (chemicals and particulates) found in the water. In turn, these physical and chemical factors support a community of biological organisms unique to a water environment.

Light

The presence and abundance of light in lakes control many biological processes. Green plants convert the light energy of the Sun into chemical energy (and ultimately plant tissue) through a process called photosynthesis. As sunlight strikes water, it is reflected from the surface (much like a mirror), scattered by particles in the water, and absorbed by the water itself. Gradually, much of the light gets used up until there is not enough light energy remaining at depth to support plant photosynthesis.

The surface depths of a lake that receive sufficient light to allow photosynthesis make up the euphotic zone. The lower limit of the euphotic zone is approximated by the 1-percent light level, or that depth where only 1 percent of the surface sunlight remains. The depth of the euphotic may be a little as 1 meter (3 feet) in very turbid (cloudy) lakes to as much as 31 meters (100 feet) in very clear lakes.

Aquatic Communities

The shallow, nearshore waters of the lake where light penetrates all the way to the bottom is called the littoral zone. The littoral community is considered the most diverse and abundant biological community in lakes.

In the littoral community, plants (macrophytes) rooted in the sediments receive enough light to grow. Some rooted plants (emergents), such as cattails, emerge from the water surface. Other plants (floating-leaved), such as water lilies, have leaves that float on the surface. Still other plants (submergents) stay entirely submerged.

The diversity of plants and the structure they add to the littoral zone attract an abundance of aquatic life. Many fish build nests here, and young fish find protection among the plants from predators. A multitude of aquatic insects (food for many fish) live on and feed among the plants and sediments of the littoral. Turtles, frogs, and many other aquatic organisms call the littoral community home.

The zone of open, deeper water found farther out from the littoral zone is the pelagic community. Here, light is still abundant, and the waters are frequently mixed by wind. Tiny, free-floating plants and animals (plankton) live here along with cruising fish.

The deep open water beneath the pelagic and euphotic zone is the profundal zone. The lack of light does not allow plants to grow here, but many fish and tiny crustaceans may still live here. Along the bottom of the lake lies the benthic zone. A variety of bottom-dwelling organisms—catfish, mollusks, worms, and midge larvae—live in this benthic community and derive their food from the sediments.

Living in the Water

Water is a medium of extreme properties that strongly shape the nature of the organisms that can survive in it. Thus, life in the water requires special adaptations. Oxygen, plentiful in the atmosphere for land animals, is much less abundant in water. Air-breathing aquatic organisms must have specialized and efficient mechanisms, such as gills, to extract oxygen from water.

Except for benthic organisms that live on the lake or stream bottom, most aquatic organisms require some means to regulate their buoyancy so that they can remain suspended in the water. Many fish have air bladders, lightweight bones, and scales—all adaptations to increase buoyancy. Plankton may have long spines or elaborate shapes to increase their surface area which slows down their sinking rate. Because ponds or streams may dry up, many aquatic organisms can enter a resting stage during development or may aestivate, as some amphibians may do in summer drought.

On the other hand, the annual range in natural water temperatures (approximately 0 to 30°C, or 32 to 86°F in temperate areas) is much lower than the range that land plants and animals must face (-28 to 40°C, or -18 to 104°F).

Food Web

Aquatic plants and animals interact with each other through a series of interconnecting pathways called a food web. Each different level in the food web or chain is called a trophic level because each represents a different type of productivity. The schematic below illustrates the food web and microbial loop for the pelagic zone of a typical fresh-water lake. Microbes are important in enabling and sustaining nutrient cycles.

Phytoplankton (predominantly algae), like terrestrial plants, require sunlight, water, and nutrients for photosynthesis. The algae and rooted macrophytes are primary producers in the aquatic environment. By converting light energy into chemical energy via photosynthesis, they create food (energy) needed for the entire aquatic food web. As such, they are at the base of the food chain. Algal groups are organized generally by color, such as green algae, yellow-brown algae, and so on.

Zooplankton, such as the shrimp-like *Daphnia* and *Bosmina,* are the primary consumers because they eat the primary producers. Zooplankton are considered herbivores because they consume plant material and are the functional equivalent of cows or rabbits on land.

Planktivores are organisms that eat zooplankton. Aquatic organisms that are planktivorous include fish, such as minnows, small sunfish, and gizzard shad, as well as a variety of aquatic insect larvae.

The piscivores are at the top of the aquatic food web and are fish-eating fish, such as bass, pike, and walleye. Piscivores are keystone species, in that their influence may cascade down the food web, affecting other organisms in lower trophic levels. For example, if the piscivore population is too high, they could eat all the planktivores. Without the fish planktivores to eat them, the zooplankton population could increase and do a better job feeding on the algae. This would lead to an increase in lake transparency. The reverse effect can happen if too few piscivores exist, which may be a result of overfishing or poor reproduction.

The purpose of EPOBIO, a Science to Support Policy Consortium funded by the European Commission, is to realise the economic potential of sustainable resources – non-food bioproducts from agricultural and forestry feedstocks. To date, our desk studies have produced eight reports addressing a range of bioproducts and feedstocks and assessing their potential for developing biorenewables with high value and utility to society. The assessment has involved a holistic analysis of the science-based projects within a wider context of environmental impact, socio economics, regulatory frameworks and attitudes of public and policy makers. This EPOBIO process has allowed costs and benefits of each application, product and process to be defined, thereby providing a robust evidence base for strategic decisions, policies and funding.

The opportunities offered by land-based agriculture, forestry and their many applications for non-food industrial products are well recognised. Most recently, the use of lignocellulose biomass for generation of transport fuels is a much debated topic in the design of future energy production systems, again illustrating the versatility of plant raw materials for both energy and non-energy products. It is in this context that the potential of marine biomass is increasingly discussed, given the size of the resource and that more than three quarters of the surface of planet earth is covered by water. These aquatic resources, comprising both marine and fresh water habitats have immense biodiversity and immense potential for providing sustainable benefits to all nations of the world. Some 80% of the world's living organisms are found in aquatic ecosystems.

Of net primary production of biomass, it is generally accepted that 50% is terrestrial and 50% aquatic. Policies of Governments have focussed almost exclusively on the use of land plants, with little consideration so far of the non-food applications and utility of macro- and micro-algae and their products. The limitations of agricultural land and the impacts of global climate change on agricultural productivity are factors of increasing relevance in the decisions that must be taken on land use for food, feed, chemicals and energy. Clearly, this increasing competition for land is driving the current consideration of the potential of the aquatic environment for the production of biofuels and industrial feedstocks.

The technical potential of micro-algae for greenhouse gas abatement has been recognised for many years, given their ability to use carbon dioxide and the possibility of their achieving higher productivities than land-based crops. Biofuel production from these marine resources, whether use of biomass or the potential of some species to produce high levels of oil, is now an increasing discussion topic. There are multiple claims in this sector but the use of micro-algae as an energy production system is likely to have to be combined with waste water treatment and co-production of high value products for an economic process to be achieved. These current biofuel discussions illustrate two issues. First, the potential broad utility of these organisms that are capable of multiple products, ranging from energy, chemicals and materials to applications in carbon sequestration and wastewater remediation. Second, the need for a robust evidence base of factual information to validate decisions for the strategic development of algae and to counter those claims made on a solely speculative basis to support commercial investment.

The current regulatory framework under development in Europe notes that an all embracing maritime policy should aim at growth and more and better jobs, helping to develop a strong, growing, competitive and sustainable maritime economy in harmony with the marine environment. An aim is to integrate existing and future EU, regional and national policies affecting marine issues. The emphasis of the proposed framework is on use of the marine environment at a level that is sustainable where marine species and habitats are protected, human induced decline of biodiversity is prevented and diverse biological components function in balance. Whilst it is recognised that innovation may help to find solutions to issues such as energy and climate change, there is little in policy proposals that addresses the utilisation of available marine biomass.

The report explores opportunities for energy and non-energy products, encompassing both marine and fresh water macro- and micro-algae. Salt water agriculture and the use of tidal flats is not discussed nor is the harvesting of aquatic plants other than algae. The first chapters briefly introduce the range of organisms and their habitats, together with the production systems that are already in development and use for their large-scale cultivation. The later chapters summarise the diverse range of products that have arisen or could be developed in this sector, including the utility of genes.

The report concludes that the macro- and micro-algal populations of the aquatic environments provide a vast genetic resource and biodiversity. This feature alone suggests that these organisms have considerable potential for offering new chemicals, materials and bioactive compounds. The completion of the genome sequencing programmes of two micro-algae also opens up major opportunities for new applications, either using the algae themselves or through using the genes in other production systems, whether fermenter-based or fields.

The culture of micro-algae has been studied widely through their potential for greenhouse gas abatement and this information is detailed in many reviews cited in this report. There are many conflicting statements on the potential of micro-algae for high biomass production, but there is a general agreement that the current production systems are not economically viable for biomass production alone. The difficulties include high capital infrastructure costs, problems of contamination through open pond systems and costs associated with harvesting and drying. These costs adversely affect the competitiveness of aquatic biomass production systems, compared to land-based agriculture and forestry.

Thus these negative cost considerations currently preclude the widespread use of micro-algae for biofuel production or production of other forms of bioenergy. Similarly, the macro-algal seaweeds, whilst used for some specialised applications, are also expensive to farm and harvest offshore. There are few clear drivers for using these species as biomass for bioenergy, except in specific circumstances such as maritime communities with no access to productive agricultural land or alternative energy sources.

The increasing concerns of global climate change and rising levels of atmospheric carbon dioxide have led to the recognition that carbon sequestration alone can have a tangible economic value. The value placed on a tonne of carbon within current trading schemes will determine decisions on how best to cost effectively 'manufacture' this product. There may be conditions in the future that would support the use of aquatic and particularly marine organisms for carbon capture and income generated through this route.

Additional value products from the micro-algae, such as chemicals, can increase the cost competitiveness. Often these are manufactured by the cells following a stress shock and under low nutrient conditions. For example, there has been a study in which the production of astaxanthin has been shown to be commercially viable using a micro-algal innoculum established in photobioreactors and transferred to open ponds for three day cultivation of biomass prior to harvest of product. This system successfully avoided the problems of contamination found in open pond cultivation systems since the cycle was extremely short.

Using micro-algae for waste water treatment is not a new idea. However, combining the ability of the cells to remediate water with their use for carbon sequestration or energy production may offer an economically viable way forward for the development of multiple products.

This report does highlight the need to consider carefully the economics of using organisms of the aquatic environment for industrial production.

Index

B

C

D

E

F

❑❑❑